International Conference on Manufacturing, Optimization, Industrial and Material Engineering

(MOIME 2014)

IOP Conference Series: Materials Science and Engineering Volume 58

Jakarta, Indonesia
29 – 31 March 2014

Editors:

Ford Lumban Gaol
Jun Ding
Jeff Webb

ISBN: 978-1-63439-869-5
ISSN: 1757-8981

Printed from e-media with permission by:

Curran Associates, Inc.
57 Morehouse Lane
Red Hook, NY 12571

proceedings.com

Some format issues inherent in the e-media version may also appear in this print version.

Printed by Curran Associates, Inc. (2015)

For permission requests, please contact the Institute of Physics
at the address below.

Institute of Physics
Dirac House, Temple Back
Bristol BS1 6BE UK

Phone: 44 1 17 929 7481
Fax: 44 1 17 920 0979

techtracking@iop.org

Additional copies of this publication are available from:

Curran Associates, Inc.
57 Morehouse Lane
Red Hook, NY 12571 USA
Phone: 845-758-0400
Fax: 845-758-2634
Email: curran@proceedings.com
Web: www.proceedings.com

International Conference on Manufacturing, Optimization, Industrial and Material Engineering

(MOIME 2014)

IOP Conference Series: Materials Science and Engineering Volume 58

Jakarta, Indonesia
29 – 31 March 2014

ISBN: 978-1-63439-869-5
ISSN: 1757-8981

IOP Institute of Physics

TABLE OF CONTENTS

IOP Conf. Series: Materials Science and Engineering **58** (2014) 011001 doi:10.1088/1757-899X/58/1/011001

Preface

The 2nd International Conference on Manufacturing, Optimization, Industrial and Material Engineering 2014 (MOIME 2014), was held at the Grand Mercure Harmoni, Opal Room 3rd Floor, Jakarta, Indonesia, during 29–30 March 2014.

The MOIME 2014 conference is designed to bring together researchers, engineers and scientists in the domain of interest from around the world. MOIME 2014 is placed on promoting interaction between the theoretical, experimental, and applied communities, so that a high level exchange is achieved in new and emerging areas within Material Engineering, Industrial Engineering and all areas that relate to Optimization.

We would like to express our sincere gratitude to all in the Technical Program Committee who have reviewed the papers and developed a very interesting Conference Program as well as the invited and plenary speakers.

This year, we received 97 papers and after rigorous review, 24 papers were accepted.

The participants come from 7 countries. There are 4 (four) parallel session and 2 Invited Speakers and one workshop.

It is an honour to present this volume of *IOP Conference Series: Materials Science and Engineering* (MSE) and we deeply thank the authors for their enthusiastic and high-grade contributions.

Finally, we would like to thank the conference chairmen, the members of the steering committee, the organizing committee, the organizing secretariat and the financial support from the conference sponsors that allowed the success of MOIME 2014.

The Editors of the MOIME 2014 Proceedings Editors

Dr Ford Lumban Gaol
Jeff Webb, PhD
Professor Jun Ding, PhD

Invited and Plenary Speakers

Keynote Speakers

1. Jeff Webb, PhD

Short Bio:
Jeff Webb is a physicist working at the University of Nottingham UK. He has a wide range of publications and research experience in several countries. His nationality is British and he holds a PhD in photothermal physics from Strathclyde University, Scotland, UK. His current research interests are the theory and modelling of ferroelectric materials and graphene-based materials.

Title:
A Two-Level Model of Rise Time in Quantum Cascade Laser Materials Applied to 5 Micron, 9 Micron And Terahertz Wavelengths

Abstract:
A two-level rate equation model for quantum cascade laser (QCL) materials is used to study the turn on delay and rise time for three QCLs with 5 micron, 9 micron and terahertz wavelengths. In order to do this it is necessary to that the model can deal with large signal responses and not be restricted to small signal responses; the model used here is capable of this. The effect of varying some of the characteristic times in the model is also investigated. The comparison of the terahertz wave QCL with the others is particularly important given the increased interest in terahertz sources which have a large range of important applications, such as in medical imaging.

2. Prof. Jun Ding, PhD

Short Bio:

Dr Jun Ding obtained PhD from Ruhr-University Bochum, Germany in 1990. After working as Research Fellow and Senior Research Fellow at University of Western Australia for 6 years, Dr Ding joined National University of Singapore as a Faculty (NUS). Now, he is Professor at Department of Materials Science & Engineering (NUS) and Visiting Professor at NIMTE, Chinese Academy of Science. Dr Ding has been working in magnetic materials for 25 years. He has published over 300 papers with > 7000 citation and H-index = 47. His current research work focuses on nanomagnetics and magnetic nanoparticles for bio-medical applications.

Title:

Synthesis of Fe3O4 nanoparticles for biomedical applications

Abstract:

Magnetic nanoparticles are promising for bio-medical applications, including MRI contrast enhancement, hyperthermia, drug delivery and separation. Superparamagnetic nanoparticles are widely used in order to avoid undesired agglomeration. However, low saturation magnetization and low magnetic susceptibility of superparamagnetic nanoparticles limit their performance. Recently, we have successfully fabricated magnetic nanorings which can be well suspended in water with suitable surfactant. Compared to superparamagnetic nanoparticles, these Fe3O4 nanorings exhibited much higher magnetization and susceptibility, so that a relatively low magnetic field is able to achieve a high magnetic flux. High R2* values have been obtained, demonstrating their great potential for T2 MRI agent. More recently, we have tested hyperthermia performance of Fe3O4 nanorings and nanodiscs. The SAR values (the measure of hyperthermia performance) obtained are much higher compared to those for reference (Fe3O4 nanoparticles). Our calculation for hyperthermia also confirms large SAR values because of their Neel and hysteresis contribution. Our work has shown the great potential of non-superparamagnetic nanorings and nanodiscs.

IOP Conf. Series: Materials Science and Engineering **58** (2014) 011002 doi:10.1088/1757-899X/58/1/011002

Workshop

Title: Optimization with Wolfram Mathematica

Speaker: Ruben Garcia

Affiliation: Mathworks

Abstract: In this workshop, there are many *Mathematica* Functionalities to be learned

IOP Conf. Series: Materials Science and Engineering **58** (2014) 011003 doi:10.1088/1757-899X/58/1/011003

Photos

IOP Conf. Series: Materials Science and Engineering 58 (2014) 011003 doi:10.1088/1757-899X/58/1/011003

Photo session part 2

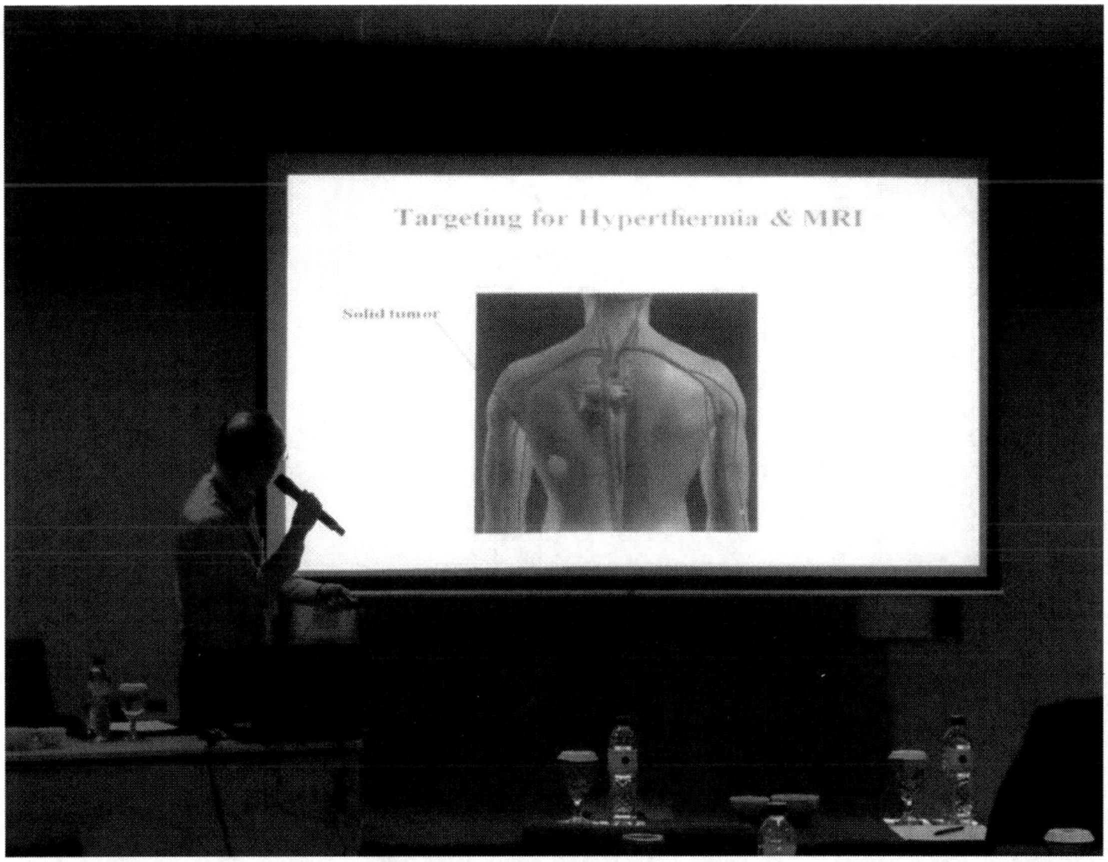

Professor Jun Ding, PhD keynote

IOP Conf. Series: Materials Science and Engineering **58** (2014) 011003 doi:10.1088/1757-899X/58/1/011003

Professor Jeffrey Webb's keynote address

The Grand Mercure Harmoni, Opal Room 3rd Floor, Jakarta, Indonesia
29–30 March 2014

Peer review statement

All papers published in this volume of *IOP Conference Series: Materials Science and Engineering* have been peer reviewed through processes administered by the proceedings Editors. Reviews were conducted by expert referees to the professional and scientific standards expected of a proceedings journal published by IOP Publishing.

Effects of Properties on S45C Carbon Steel by Electroless Ni-P Adding Al_2O_3 Powder of Composite Deposition and Various Heat Treatments

Shih-Hsien Chang*, Cheng Liang and Chih-Chung Chang
Department of Materials and Mineral Resources Engineering, National Taipei University of Technology, Taipei 10608, Taiwan, R. O. China

E-mail: changsh@ntut.edu.tw

Abstract. This research studies the effects on the surface microstructure and mechanical properties of S45C carbon steel *via* electroless Ni-P, by adding αAl_2O_3 powder via composite deposition and various heat treatments. The plating specimens were treated with pH5 and pH8 baths heated to 350°C and 300°C, respectively, and soaked for 1 hr. Meanwhile, two different particle sizes of Al_2O_3 powder were added to the electroless Ni-P plating: 0.3 μm of αAl_2O_3 and 0.05 μm of γAl_2O_3 powder, respectively. The experimental results show that, after 1 hr of heat treatment at 300°C, the optimal hardness for the specimens using the pH8 of Ni-P with αAl_2O_3 and γAl_2O_3 added by composite deposition are $HV_{0.05}$ 1237 and $HV_{0.05}$ 1145, respectively. All of the specimens underwent the main precipitate phase of Ni_3P after heat treatment.

1. Introduction

Electroless Ni-P deposits have been used extensively in many industries due to their wear and corrosion resistance [1,2]. The properties and structures of the Ni-P coatings depend on its phosphorus content. After suitable heat treatment, the Ni-P coating could be strengthened with the crystallization of nickel and nickel phosphide [4,5]. S45C carbon steel (S45C) has been widely used in industrial applications, such as crankshafts, gears, main spindles of machine tools, connecting rods, etc., because of its distinguished mechanical property [6,7]. Several surface hardening for S45C have been carried out and discussed [8]. Moreover, composite coatings constitute is a new class of materials which are mostly used for mechanical and tribological applications [9,10]. The aim of this study was to evaluate the surface microstructure, wear resistance and phase transformation *via* various heat treatments of electroless Ni-P adding αAl_2O_3 and γAl_2O_3 composite depositions in regard to S45C carbon steel.

2. Experimental Procedures

In this study, JIS S45C carbon steel was used as a substrate, the specimen size is Ø36 × D5 mm; chemical compositions are listed in Table 1. Furthermore, the parent of S45C is oil quenched at 850°C, tempered at 400°C, and the tempering process is repeated twice to reach a hardness of $HV_{0.05}$ 421. In addition, the plating bath is divided into an acid and alkaline bath (pH5 and pH8); sulfuric acid and ammonia were used to adjust the pH value. The bath compositions, concentration and operational conditions are listed in Table 2. As reported in our previous study, the optimal pH values of electroless Ni-P plating for S45C carbon steel were pH5 and pH8 when soaked for 1 hour at the optimal heat treatment temperatures of 350°C and 300°C, respectively. In order to further improve the surface hardness and wear of the samples, electroless Ni-P plating, with Al_2O_3 powder added by composite deposition, with different particle sizes of αAl_2O_3 and γAl_2O_3 powders, was prepared. After the electroless plating process, the pH5 and pH8 baths specimens were heat treated at 350°C and 300°C, respectively, and soaked for 1 hr. The vacuum was kept at a 10^{-1} atmosphere and the heating rate was 10°C min^{-1}.

Hardness of the specimens was measured by Vickers indenter (HV) with loading of 50 g, which complied with the ASTM E384-08a standard method. The wear resistance of specimens was evaluated in a ball-on-disk test. The wear parameters are: specimen size is Ø36 × D5 mm, diameter of Cr ball is 6 mm, axial load is 4.9 N, disc rotation is 100 rpm, sliding speed is 0.031 ms^{-1}, and total rotation is 3000 revolutions. The coefficient of friction was obtained from POD-FM406 software and the wear depth was measured by α-step equipment (ET- 4000A)

Table 1 Chemical composition of the S45C carbon steel (mass %)

C	Si	Mn	S	P	Fe
0.46	0.21	0.72	≦0.04	≦0.03	Bal.

Table 2 Bath composition and operating conditions of S45C carbon steel by electroless Ni-P adding Al_2O_3 composite deposition

Process	Bath composition	Concentration (g/L)	Condition
Acid bath	NiSO4 · 6H2O	30	84±1°C
	NaH2PO2 · H2O	15	1 hour
	Na2C4H4O4 · 6H2O	12	pH 5
Alkaline bath	NiSO4 · 6H2O	30	84±1°C
	NaH2PO2 · H2O	10	1 hour
	NaC6H5O7 · 2H2O	40	pH 8

3. Results and Discussion

Fig. 1 shows the cross-section morphology of pH5 and pH8 on electroless Ni-P with the addition of 0.3 μm of αAl_2O_3 and 0.05 μm of γAl_2O_3 via composite deposition, respectively. It could be clearly seen that the coating containing the Al_2O_3 particles, despite the effect of a slight agglomeration phenomenon, showed a good overall dispersion on the samples. Table 3 shows the composition analysis of the cross-section of the pH5 and pH8 electroless Ni-P composite depositions. Obviously, the γAl_2O_3 content was less than that of αAl_2O_3, which could have been the reason why the smaller particles easily underwent the agglomeration phenomenon. Another possible factor was that the effect of the mixing made it easier for the smaller and lighter particles of the γAl_2O_3 powder to form a suspension, but harder for their co-deposition with Ni-P plating. Consequently, the γAl_2O_3 was not as effective in adsorbing the Ni-P coating, and thus its content was lower. Further comparison with the our previous study showed that the thickness of electroless Ni-P with the addition of the Al_2O_3-powder coating was slightly less than without it, as the nickel-phosphorus in the plating process must co-precipitate with the Al_2O_3 powders in order for the result to be a thinner coating.

Fig. 2 shows the cross-section morphology of pH5 and pH8 on electroless Ni-P with αAl_2O_3 and γAl_2O_3 added by composite depositions after heat treatment, respectively. Figs. 4c and 4d represent the cross-section of Ni-P with the addition of 0.05 μm of γAl_2O_3 by composite deposition; clearly, the agglomeration phenomenon was more severe than when αAl_2O_3 powder was added (as seen in Figs. 2a and 2b). It seems reasonable to conclude that the relatively large surface energy of the fine γAl_2O_3 powders resulted in the larger attraction; thus, aggregates were easily produced as the overall adsorptive capacity of αAl_2O_3 was greater than that of γAl_2O_3. In addition, the extra-fine γAl_2O_3 powder was easily affected by the stirring in the plating bath, which accounted for it not being adsorbed during the composite deposition, and then immediately being removed by the stirring bath.

Table 4 shows the surface hardness of pH5 and pH8 on electroless Ni-P with added αAl_2O_3 and γAl_2O_3 by composite deposition both without and with different heat treatments. In this study, the particle size (αAl_2O_3 or γAl_2O_3) did not obviously affect the surface hardness of the electroless Ni-P composite depositions; however, the pH value and heat treatment temperature significantly affected the surface hardness. The pH8 of Ni-P with added αAl_2O_3 or γAl_2O_3 powder had a greater hardness than those of the pH5 of Ni-P composite deposition. The greatest hardness was $HV_{0.05}$ 927, which appeared in the pH8 of Ni-P with added αAl_2O_3 by composite deposition. At the same pH value, the αAl_2O_3 powder had better dispersion than that of γAl_2O_3, which resulted in a greater hardness. Previous literature [6] also indicated that a bigger particle size results in better dispersion and greater hardness after the electroless plating treatment, which also agrees with our findings.

In this study, all the specimens of Ni-P with added αAl_2O_3 and γAl_2O_3 by composite deposition had greater hardness after heat treatment. The main reason could be ascribed to the phase transformation of the Ni-P depositions. As a result, the surface hardness of the Ni-P composite depositions obviously increased after heat treatment. Furthermore, the greatest hardness was $HV_{0.05}$ 1237, which appeared in the pH8 of the Ni-P with added αAl_2O_3 by composite deposition after a 300°C heat treatment. The hardness of the pH8 of Ni-P with γAl_2O_3 added by composite deposition was $HV_{0.05}$ 1145 after a 300°C heat treatment. It was confirmed that the adding of either αAl_2O_3 or γAl_2O_3 powder to the electroless Ni-P deposition significantly increased the surface hardness of the Ni-P plating on S45C carbon steel after suitable heat treatment. These results suggest that both the better dispersion of Al_2O_3 and the phase transformation of the Ni-P deposition play important roles in improving the surface hardness of Ni-P plating on S45C carbon steel.

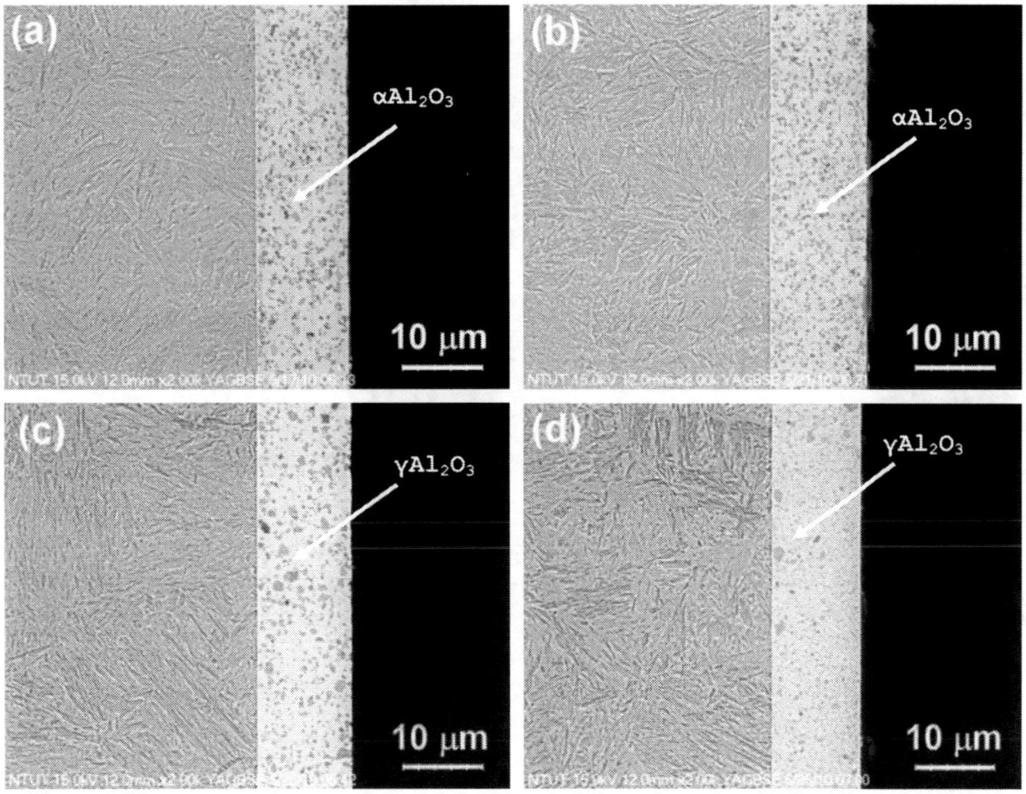

Figure 1 The cross-section morphology of pH5 and pH8 on electroless Ni-P adding αAl_2O_3 and γAl_2O_3 composite depositions (a) pH5 αAl_2O_3, (b) pH8 αAl_2O_3, (c) pH5 γAl_2O_3, (d) pH8 γAl_2O_3.

Table 3 The composition analysis of the cross-section of pH5 and pH8 on electroless Ni-P adding Al_2O_3 composite depositions (mass %)

Powder	pH value	Ni	P	Al	O
αAl_2O_3	5	90.48	5.57	1.83	2.12
	8	90.59	5.02	2.15	2.24
γAl_2O_3	5	91.09	5.53	1.57	1.81
	8	91.70	5.28	1.29	1.73

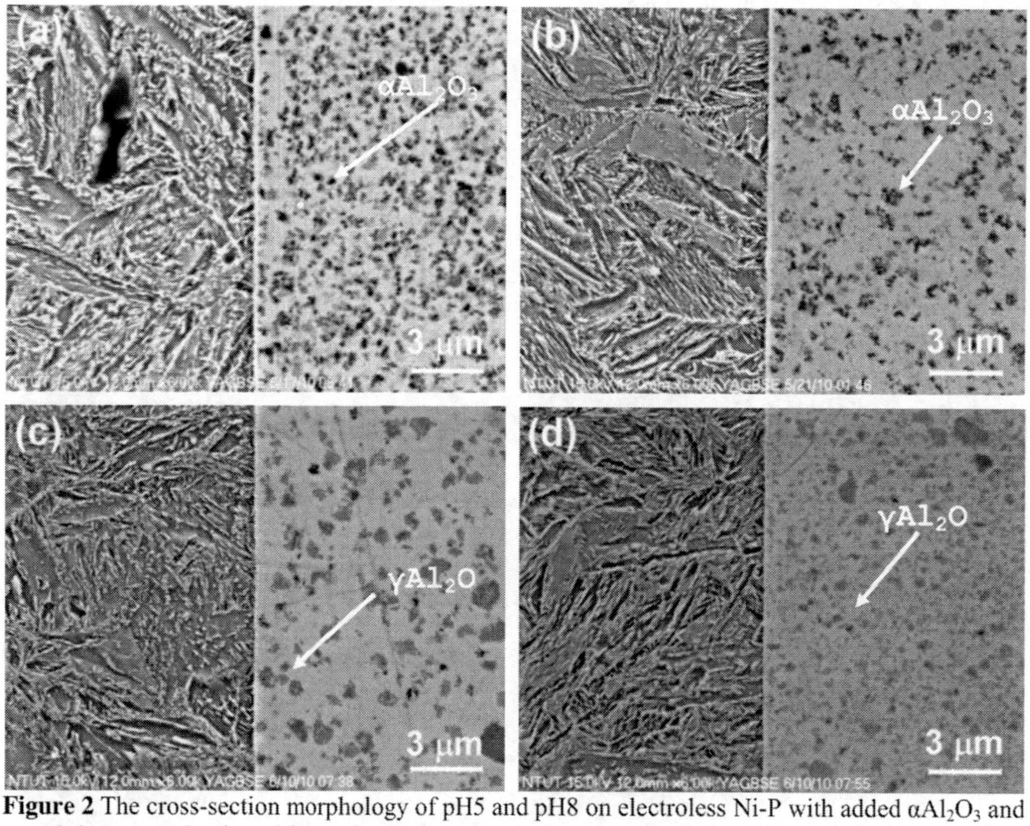

Figure 2 The cross-section morphology of pH5 and pH8 on electroless Ni-P with added αAl_2O_3 and γAl_2O_3 composite depositions after different heat treatment (a) pH5 αAl_2O_3 and 350°C, (b) pH8 αAl_2O_3 and 300°C, (c) pH5 γAl_2O_3 and 350°C, (d) pH8 γAl_2O_3 and 300°C.

Table 4 Comparison of the surface hardness of pH5 and pH8 on electroless Ni-P with added αAl_2O_3 and γAl_2O_3 composite depositions with non heat-treated and different heat treatments ($HV_{0.05}$)

Materials Condition	Hardness($HV_{0.05}$)
pH5(αAl_2O_3)	739
pH5(αAl_2O_3) after 350°C heat treatment	1095
pH8(αAl_2O_3)	927
pH8(αAl_2O_3) after 300°C heat treatment	1237
pH5(γAl_2O_3)	730
pH5(γAl_2O_3) after 350°C heat treatment	1045
pH8(γAl_2O_3)	889
pH8(γAl_2O_3) after 300°C heat treatment	1145

The wear test was used to evaluate the wear resistance of Ni-P with Al_2O_3 added by composite deposition. Fig. 3a shows the coefficient of friction of pH5 and of pH8 on electroless Ni-P with αAl_2O_3 added by composite deposition without heat treatment, or after different heat treatments. The lowest coefficients of friction were 0.411 and 0.437, which appeared in the pH5 and pH8 of Ni-P with αAl_2O_3 added by composite deposition without heat treatment, respectively. Fig. 3a also shows that, after heat treatment, the average coefficient of friction (~0.5) of Ni-P with αAl_2O_3 added by composite deposition is higher than without heat treatment. Because the composite depositions from the original amorphous or microcrystalline structure transferred to the crystalline structure of the Ni_3P phase precipitation, it was more dense than amorphous. Moreover, the effect of precipitation hardening of the Ni_3P phase and the added Al_2O_3 powder led to greater friction between the chrome steel ball and the contact surface of the sample during the sliding test. Thus, the coefficient of friction was higher after heat treatment; this result also agrees with the previous study [10-13]. Fig. 3b shows the coefficient of friction of pH5 and of pH8 on electroless Ni-P with γAl_2O_3 added by composite depositions without heat treatment or after different heat treatments, with a similar trend as in Fig. 3a. The coefficient of friction for the pH5 of Ni-P with added γAl_2O_3 was minor (0.422) without heat treatment. Significantly, the effect of the lubricant explained the lower coefficient of friction.

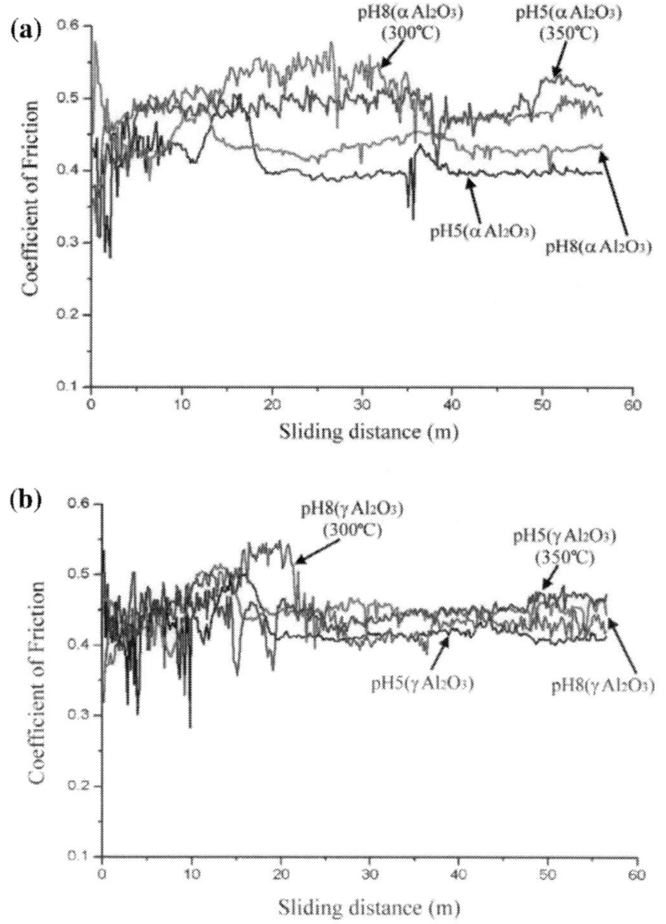

Figure 3 Comparison of the coefficient of friction of pH5 and pH8 on electroless Ni-P with added Al_2O_3 composition depositions when non heat-treated or after different heat treatments (a) αAl_2O_3 and (b) γAl_2O_3.

4. Conclusions

In this study, the hardness of the electroless Ni-P deposition was increased by adding Al_2O_3 powders. Without heat treatment, the maximum hardness was $HV_{0.05}$ 927, which appeared in the pH8 of electroless Ni-P with αAl_2O_3 added by composite deposition. The greatest hardness was $HV_{0.05}$ 1237, which appeared in the pH8 of electroless Ni-P with αAl_2O_3 added by composite deposition after 1 hr of heat treatment at 300°C. The lowest coefficient of friction (0.411) appeared in the pH5 of electroless Ni-P with αAl_2O_3 added by composite deposition without heat treatment because of the effect of the lubricant. In addition, the coefficient of friction of Ni-P with Al_2O_3 added by composite deposition was higher after heat treatment. From the wear morphology point of view, both showed obvious abrasive wear but also better wear resistance.

Acknowledgments

The authors would also like to express their appreciations for Professor Cheng Liang.

References

[1] R. Elansezhian, B. Ramamoorthy and P.K. Nair 2008 *Surf. Coat. Technol.* **203** 709-712
[2] X.G. Hu, W.J. Cai, Y.F. Xu, J.C. Wan and X.J. Sun 2009 *Sur. Eng.* **25** 361-366
[3] S.K. Tien, J.G. Duh and Y.I. Chen 2004 *Thin Solid Films.* **469-470** 333-338
[4] L.L. Wang, H.J. Chen and Z.L. Chen 2011 *Sur. Eng.* **27** 57-60
[5] Y.T. Yoo, D.G. Ahn, K.B. Ro, S.W. Song, H.J. Shin and K. Im 2004 *J. Mater. Sci.* **39** 6117-6119
[6] Y. Chino, K. Shimojima, H. Hosokawa, Y. Yamada, C. Wen, H. Iwasaki and M. Mabuchi 2002 *J. Mater. Sci. Lett.* **21** 1695-1697
[7] S.H. Chang, T.P Tang and F.C. Tai 2011 *Sur. Eng.* **27** 581-586
[8] H. Matsubara, T. Yonekawa, Y. Ishino, N. Saito, H. Nishiyama and Y. Inoue 2006 *Electrochimica Acta.* **52** 402-407
[9] S.H. Chang, K.T. Huang and Y.H. Wang 2012 *Materials Transactions.* **53** 745-751.
[10] S.H. Chang, T.P. Tang, K.T. Huang and J.K. Chen 2012 *ISIJ International.* **52** 499-504.
[11] V.V.N. Reddy, B. Ramamoorthy and P.K. Nair 2000 *Wear.* **239** 111-116
[12] S.H. Chang, C.C. Chang and C. Liang. 2013 *IOP Conference Series: Materials Science and Engineering (MSE).* **46** 012003 1-6
[13] S.H. Chang, Y.K. Lin and K.T. Huang 2012 *Surf. Coat. Technol.* **207** 571-578

Smart Equipment and Virtual Resources trigger Network Principles in Manufacturing

H Kuehnle

Otto-von-Guericke-University Magdeburg, IAF, Universitaetsplatz 2, D 39106 Magdeburg, Germany

E-Mail: hermann.kuehnle@ovgu.de

Abstract. Computing miniaturization and smart devices rapidly change manufacturing. Decentralization and atomisation of resources uncover novel manufacturing behaviour. Virtual representations of units, processes and resources enforce unaccustomed network principles with strategic impact and irrefutable implications on manufacturing. Eventually manufacturing in total might have to be reconsidered. Distributed manufacturing, largely engaging interacting intelligent manufacturing units and decentralized planning, may be regarded as the manufacturing future. Gradually evolving decision procedures clearly illustrate important effects of irreversible shifts of focus towards units' collaboration and interoperability.

1. Introduction

In manufacturing, newly available technologies offer so far unseen opportunities. Technologies for information processing and communication are about to embrace all important manufacturing areas. Real production increasingly melts with the digital production world. With novel information technology, smarter equipment and networked units, many factories gradually turn into large computing units sending data across and outside companies. Numerous approaches for computerising manufacturing units and processes propagate powerful and fascinating services, ready for implementation. The appearance of novel devices, able to be positioned, to be tracked, to be identified on one hand, also capable to communicate, to act, to negotiate and even to decide on the other is gaining influence on everything that concerns manufacturing. As manufacturing increasingly supports the processes by means of virtualized smart resources, distributed manufacturing (DM) [17] irreversibly extends from automated factory floors onto manufacturing enterprises in total. Advances on the fields of embedded systems, and cyber physical systems additionally accelerate this shift. Important developments are also telecommunication driven and discussed under different chapters, as Internet of things, Ubiquitous Computing, Cyber Physical Systems, Smart Objects or comparable terminology. Decentralisation and atomisation of processes, units and procedures and their virtualisations are in trend. Some principles that had been found for DM now reappear for manufacturing in total, so most upcoming set ups can be mirrored to the DM experiences and respective findings.

As various communities from different disciplines outside of manufacturing are intensively working on new services and novel devices, the most important developments, relevant for manufacturing, are sketched. On the respective fields, the literature, most frequently referred to, as well as pioneer papers with coining terminology are cited; recent developments in manufacturing are introduced in accordance with international, governmental or industrial institutions. The proposed virtualisations are largely based on information models, so a structured overview on frequently used models in manufacturing is recalled. To obtain closed and coherent descriptions of networks, topological spaces are introduced as a base for further discussions. The space construct, as outlined, reduces down to the essentials on one hand; on the other hand it is powerful enough to capture all relevant aspects of networked manufacturing. The possibility of smoothly attaching model worlds to the nodes of the space literally imposes interpretations of cyber physical production and smart objects. The resulting set of loosely coupled, autonomously acting manufacturing units are evidently subject to principles and modes of complex structures that are known from advanced mechatronic systems and DM set ups [18] already. In this context, procedures for controlling the behaviour of units and the generalized principle of encapsulation are outlined for generalisation. The final section is devoted to technical and managerial implications.

The paper aims at offering a more comprehensive theory base for virtualizing manufacturing which helps to anticipate an emerging "smart age" of manufacturing. Practitioners should be provided with some trajectories on these rapid developments and expected impacts as inputs for their decisions and to verify their gut feelings for the next steps in organising manufacturing network processes.

2. ICT Breakthroughs with Disruptive Impact

Intensive research and development in the fields of telecommunications, computer science and engineering shows vast progresses. Miniaturised, smart and multi-functional electronic devices with enormous computation capabilities and with high mobile, ubiquitous, uninterrupted, and embedded capabilities are already in use in daily life [1]. In consequence, ICT changes the working environment in manufacturing as well. Some important IT developments provoked intensive actions and reactions in the manufacturing world, not only on company levels but on national and international levels [28] as well. Considering the fact that telecommunication, software and computer industries involve biggest players with enormous research capacities on this field, more brilliant innovations will be ahead with radical and disruptive consequences throughout manufacturing. Their potential cannot be ignored by manufacturing companies as the use of these novel ICT achievements are expected be comfortable and efficient on various other sectors; undoubtedly they will irreversibly and broadly find their ways into the manufacturing world.

2.1. *Cyber Physical Systems*

Some years ago, an object virtualization method has emerged, known as Cyber-Physical System (CPS, also DCPS if distributed) [26], [32], [36], meaning the integration of computing systems with physical processes and physical environments. Major motivation behind the notion of CPS is the need to design and produce reliable and sustainable computing systems that work in harmony with their surroundings [30]. Components are networked at every scale and computing is deeply embedded into every physical component, possibly even into materials [33], [7]. Exploiting CPS for manufacturing has brought up the terminology of Cyber Physical Production Systems (CPPS), which is strongly propagated in the national funding scheme Industry 4.0 in Germany. The introduction of CPS into automation is expected to break up the monolithic functional automation pyramid into virtualized distributed networked nodes that are more difficult to handle; therefore machine toolmakers claim urgent need for joint actions. When using CPS [35], components could adapt themselves automatically to the other components, which inevitably changes the way in which these CPS-enabled components are designed and manufactured. Therefore manufacturers see reasons for totally rethinking industry and industrial production when establishing CPPS to take full advantage of the introduction of CPS, [35].

2.2. *Internet of Things (IoT), Pervasive Computing and Smart Object (SO)*

Some text Parallel to CPS (US), computer scientist had come up with the Internet of Things (IoT) in the context of ERA (EU). CPS and IoT cannot be clearly differentiated, since both concepts have been driven forward in parallel, although they have always been closely related [5]. The IoT is considered a part of the future internet and could be defined as a dynamic global network infrastructure with self-configuring capabilities, with physical and virtual objects carrying own identities and specific attributes like virtual personalities; they use intelligent interfaces and are online connected the information network. IoT is of interest to manufacturing. In industry, the "thing" may typically be the product itself, the equipment, the transportation means, etc. It is obvious that these developments, too, accelerate the integration of Smart Objects in the Internet. Additionally pervasive computing has migrated from desktops to mobile phones, and computing is increasingly included into a variety of objects.

By adding more data to objects, we are witnessing the upcoming of a large IoT, where every object has his proper identity (e.g. [8]). We shall experience smart worlds full of smart objects [14]. A Smart Object (SO) is an autonomous physical/digital object augmented with sensing, processing, and network capabilities. RFID technology is closely linked to SOs. In contrast to RFID tags, SOs carry chunks of application logic that let them make sense of their local situation and interact with human users. Coupled with software agent technology however, RFID can transform everyday objects into smart objects as well [6]. Therefore smart objects and all the developments around are widely considered as of highest relevance for manufacturing, because RFIDs are applied in manufacturing as solution components since years already.

2.3. Ubiquitous Computing (UC), Cloud Computing, Cloud Manufacturing (CM), Grid Manufacturing, Cloud Based Distributed Manufacturing and Hybrid Clouds

UC, too, has contributed to upgrade objects to become smart objects, which can provide new services that could not have been imagined before, because of the steady connection between the real-world objects and the intelligence of information systems. UC denotes another vision of a future world of smart objects, physical items with physical shape and function being extended into virtuality [20]. Miniaturizations of computer technology will result in smallest processors and sensors for integration into more and more everyday objects, outdating accustomed computing equipment. Instead, people will be able to communicate directly with their clothes, or watches, or pens, and these objects will communicate with each other and with other people's items [10]. UC cannot be considered a proper technology or classified as a new functionality, rather as a set of functions which as a sum creates novel computing [31].

Cloud architecture may provide users with computing resource options of configurable, virtual manufacturing networks, based on models like federated factories or software service applications [24]. Cloud computing is a novel model for enabling ubiquitous on-demand network access to a network of computing resources that can be instantly accessed and released with little effort or third party interaction [26]. A cloud is parallel and distributed computing net, composed of a set of interconnected computers, presented as one unified computing resource that is available according to service-level agreements [3]. Virtualisations of resources and fast interconnections open up companies in general and manufacturing areas in particular to new services and services' architecture i.e., cloud hardware-as-a-service (HaaS), cloud software-as-a-service (SaaS), cloud platform-as-a-service (PaaS), cloud infrastructure-as-a-service (IaaS). Virtualized computing resources allow big data storage and rpocessing, cloud ERPs and Cloud CRMs are already available, online monitoring and positioning of all products and resources enable tracking and fixing issues in real time, allowing companies to instantly improve all attributes of their manufacturing process."

A number of researchers already propagate to specify Cloud Computing into Cloud Manufacturing CM (e.g. [40]). CM is anticipated as a new mode of networked intelligent manufacturing which is service-oriented, highly efficient and advocates immediate implementations of the concept of Cloud Computing in manufacturing. CM may, indeed, become a networked manufacturing mode with quickest responses to market demand, enhanced competitiveness and facilitated collaborative manufacturing [41]. Furthermore Resource Cloud Encapsulation RCE of soft and hard manufacturing resources with resource sharing is projected as services for further resource virtualization in CM [25].

In IoT and CPS, information technologies are used to access and to connect manufacturing resources. In RCE, moreover, all physical manufacturing resources are seen as transferred into logical services; based on complete resource virtualization, RCE technology constructs large-scale virtual manufacturing resource pools that can be used for interacting and feedback control of manufacturing. RCE is supposed to largely reduce the coupling between physical resource and manufacturing application by the transferring physical resources into logical resources and virtual CM services with instant utilization, high agility, high security and high reliability. In addition, resource pooling and virtualization might enable even more sophisticated solutions under Cloud-Based Design and Manufacturing (CBDM). This is a type of parallel and distributed system consisting of a collection of inter-connected physical and virtualized service pools of design and manufacturing resources [38] possibly triggering new perceptions of product design and manufacturing.

All cloud solutions enable to dynamically adapt in order to satisfy unpredictable or unexpected demand. Such manufacturing clouds support scalability to a certain level, e.g. manufacturing units, general purpose equipment, and standard components for machining. Given that the cloud is a huge shared service pool of design and manufacturing resources, it may also be possible for cloud service consumers to find some dedicated tools and equipment for some specific products available in the manufacturing cloud that can satisfy their requirements [37].

Public clouds are handled by external providers, and the data of various clients may be mixed in factories, servers or CPUs of the cloud. End users do not know, if other clients' orders are executed, too, in a factory, not even in one and the same manufacturing unit. Private clouds are a good choice for companies that feel the need for high data protection. Hybrid clouds that combine the models of public and private clouds may be the key to achieving an external supply in scale form and under demand, but these clouds add the complexity of determining how to allocate tasks and processes across these different environments [23].

3. Model World of Manufacturing and ICT trajectories

Whenever we talk about interacting, negotiating and communicating objects, we always talk about respective models of these objects performing such activities. Also planning, decision and execution in manufacturing does obviously not regard the units themselves but certain models and attributes of these units that configure and are put into relations. Each step may make use of a number of models interacting, raising the question of how their dependencies and simultaneous actions influence choices, highlight attributes or require certain levels of detail of these models to be involved. The manufacturing network units' interaction structure must be envisioned as an interrelations' structure of specific models, representing these units. Envisioned like this, manufacturing does not just consist of simple units but of objects that encapsulate rich model structures, able to unfold numerous attributes and properties into the attached realm of models.

Manufacturing networks may then be interpreted as specific Hausdorff spaces. The topological nature of Hausdorff spaces allows identifying network units (nodes) and to attach tangent spaces to each one [19]. The set-up is rich enough to capture a vast majority of configurations and decision situations occurring in manufacturing networks. This is accomplished by "attaching" tangent spaces carrying adequate models, attributes, relations and aspects assigned to the manufacturing networks' nodes (Fig. 1). Moreover, these virtualizations of manufacturing objects, also called mappings, capture e.g. encapsulations of behaviour, fold and unfold properties, on-off modes of self-organization. Configurations may be mapped and monitored as well by models, indicators and attributes, and the views are expressed by composite attached models, the reason why all mappings are assumed to be homeomorphous.

In practical terms, the homomorphism postulate stands for compatibility of models of different units. Models of tasks of different units can form a process flow model only and models of machines of different units compose a useful layout only, if the respective units' models are compatible. To be able to do this easily, all involved virtualizations of the units will have to be standardised in some way, so a collection of units represented by attached models is instantly able to link, to interact and to execute important procedures e.g. for manufacturing planning, structuring, operating, linking, improving and deciding.

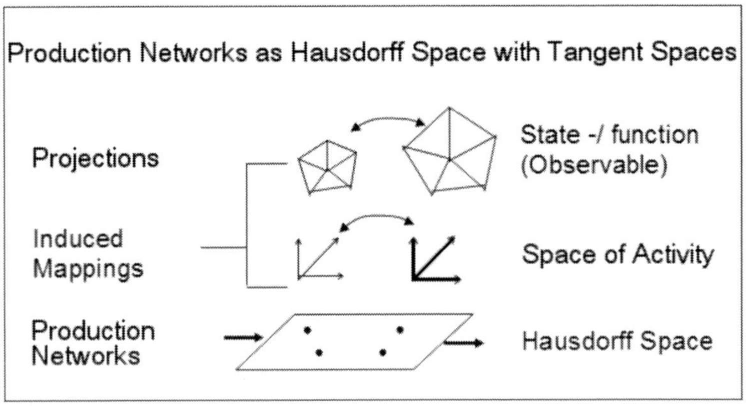

Figure 1. Production Networks' units with attached tangent spaces (models) as mappings of Hausdorff Space nodes according to [16]

3.1. Models for manufacturing management and -planning

To answer the question about which models are to be attached for manufacturing applications, which properties and attributes ought to be mapped, the chapters of manufacturing systems planning and control history may be recalled.

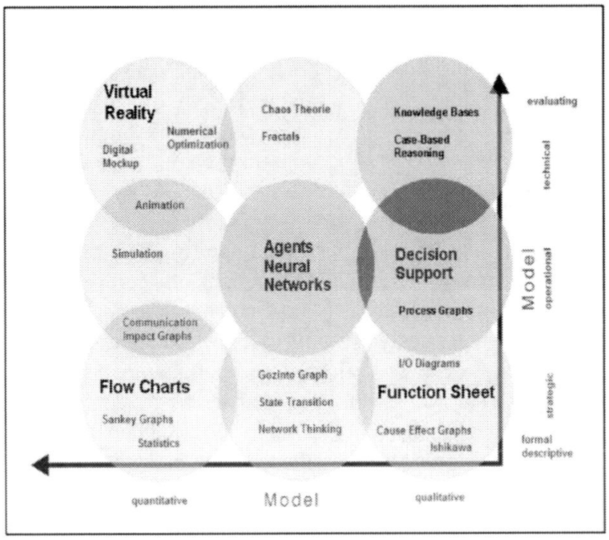

Figure 2. Portfolio of generally used information models for manufacturing planning and decisions, attachable to the network units according to [18]

With the sophistication of manufacturing, important abstractions and experiences have been consolidated into a collection of generally recognised models, instruments and tools. With the introduction of computers in manufacturing, many of these models and instruments (or derivates thereof) have been successfully incorporated in standard software e.g. ERP, Cave, DSS or facilities' planners (Figure 2). Manufacturing management generally makes intensive use of these models and model systems for specific problem solving, routine decisions and planning support, for instance for shopfloor planning, adequate models are flow charts, Sankey graphs, DMU/VR based on geometry data of buildings and machines. A structured approach to display important sets of models for manufacturing has been made in the context of concurrent enterprising and collaborative distributed planning, which could be considered as a base for virtualizing manufacturing networks.

3.2. ICT use and Manufacturing Models' availabilities

Of course, the application of the models, as described above, depends on the availability of computing power and respective software. Even for the near future it is foreseeable that most smart units in manufacturing will have enough computing power to carry these as an encapsulation. Just to trigger some ideas, the progress in decentralized ICT support with availability of these models for manufacturing processes may be illustrated over the time line and the data volumes involved (Figure 3).

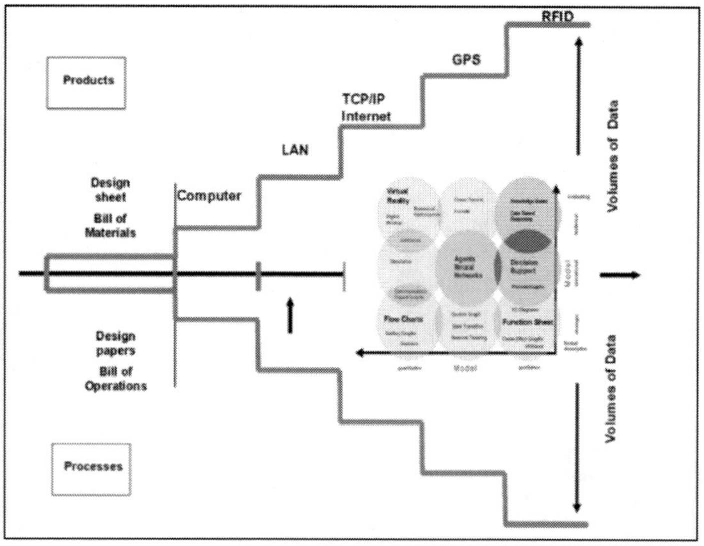

Figure 3. ICT devices and corresponding data volumes and respective model worlds to be implemented, arrayed on a timescale.

As one trajectory for future manufacturing it may be kept in mind, that manufacturing units may be imagined as carrying all the models discussed above ready for application to link, to compose, to negotiate and to decide, processed by own computing power or remote. Manufacturing then appears as a set of loosely coupled autonomous smart units, spontaneously forming networks and executing processes; concurrent and evolving planning; negotiating decisions, all by interrelating models. This appearing set up seems to be quite different from what we are accustomed to when describing manufacturing and manufacturing management. Therefore it may be considered worthwhile to take a closer look at this emerging world of smart manufacturing units and the rules of the game there and to search for characteristics and principles.

4. Principles, Properties and Modes

On a smaller scale, many of the phenomena stated have already been encountered with configurations of distributed manufacturing [17]. Seeing all the similarities, the attempt to generalise and widen up important principles that have been identified for loosely coupled manufacturing systems, to the manufacturing networks' level, appears most promising. For start, a list of properties for smart units in manufacturing may be given that support compounding manufacturing processes of networked elements. Most of the capabilities, which smart objects for general use include already, are suitable for manufacturing process set ups, therefore their adaptation is less a question of requirements fulfilment, and it seems to be more a specification matter.

4.1. Smart Units' Properties and Requirements

4.1.1. Modularity. For increasing flexibility of operations as well as the ability to be easily reconfigured due to changing conditions, typically modularity is introduced into manufacturing operations and equipment design. Modularity is generally followed by distribution of functionalities, frequently accompanied by physical or/and geographical distribution. The principle of modularity is known and widely used in manufacturing and organisation already.

4.1.2. Heterogeneity. Due to the variety of devices and units engaged, the manufacturing networks are intrinsically heterogeneous. Heterogeneity can occur on various levels and for a number of reasons; on the technical level, heterogeneity originates from different hardware platforms, operating systems, database management or programming languages; on the conceptual level, heterogeneity originates from differences in understanding and modelling the same real-world phenomena.

4.1.3. Time Synchronisation. Timekeeping technologies, e.g. from positioning system satellites and the Network Time Protocol provide real-time approximation of Coordinated Universal Time (UTC, world time standard) are used.

4.1.4. Interoperability. Technologies for realizing smart devices have already been around for years, standardized by the IETF. Interoperability is the ability of two or more systems or components to exchange information and to use the information that has been exchanged [12], [15]. In distributed manufacturing, interoperation abilities are aimed at on all levels, people resources, between enterprises and enterprise units.

4.1.5. Scalability. The capability to extend resources in a way that no major changes in structure or application are necessary is generally referred to as scalability. Due to stronger links between cyber objects and real manufacturing units, the term of scalability becomes relevant for manufacturing systems too. For example, the cloud system allows the cloud service consumers to quickly search for and fully utilize resources, such as idle machines, in a different company to upscale manufacturing capacities.

4.2. Concurrency Modes and Mechanisms

Recognizing these potentials is surely not exaggerated to postulate the necessity of a complete re-thinking of manufacturing and a thorough revision of every well established and habitually used, so far proven and uncontested, manufacturing setup. It's not only the fact that all solutions have been set up without employing such options and technical possibilities, it is no longer possible to establish factory centred solutions on the base of pure systems thinking, widely ignoring the network nature of manufacturing. Most prominent examples are deeply rooted for example the term of process and supply chain in manufacturing; in reality we work on the base of process and transformation stage networks, which expose process chains ex post as planning and decision results.

4.2.1. Behaviour. Behaviour is the range of actions made by systems, or abstract units, in interaction with other units and the environment. A unit shows its state in indicators (variables, data) and exposes its behaviour through methods (functions) that react to certain events. Process parameters present the behaviour of a unit and its interactions with other objects. Monitoring tools enable the users to specify and to process-level events such as inter process communication, as long as these events are at the correct level of abstraction of the network units, as successfully applied in DM [17]. As a representation of the units' behaviour, Spaces of Activity (SoA) may be described by the units' objectives, the resources and constraints. In consequence, the SoA volume may be identified as the unit's decision space i.e. admitted zone for the units' state (Figure 4). The unit's behaviour, e.g. expressed by corresponding indicators, gives input for decisions on maintaining the unit's self-organization mode or reducing autonomy and calling for external interference. In cases of a unit's inability to cope with the objectives or the changes in the environment, network "order parameters" may gain influence on the units' activities ((self) reproduction, (self) destruction, (self) structuring).

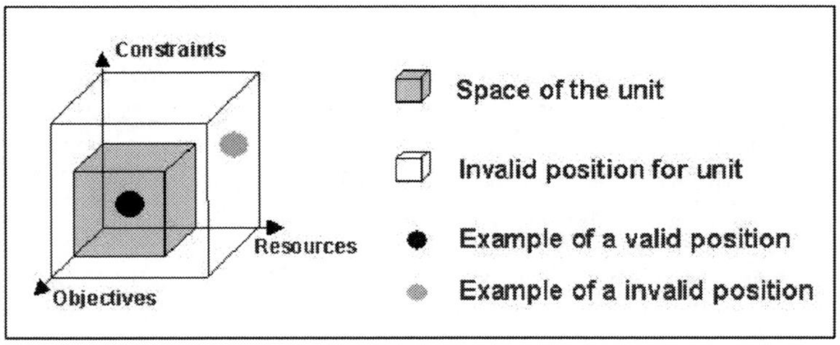

Figure 4. The Space of Activity (SoA), a model mapping of network nodes, monitoring the behaviour of the unit by relevant indicators and observable.

This "biologically" [34] as well as "fractal geometric" [xx] inspired manufacturing model approach addresses challenges in complex (unpredictable) manufacturing environments linked to self-organization, restructuring and adaptation. It makes easily adaptable to changes in manufacturing environments, and coins the global behaviour by the effects of interaction between units [2]. Applied for manufacturing network decisions, such behaviour thinking supports the often encountered "levelled" manufacturing network adaptation procedures.

4.2.2. Parallelism. An optimum base for collaborating using least resources and time is to do substantial steps towards parallelism of all actions and operations. Parallelism aims at reducing execution time or improving throughput. Adding parallelism to an event driven view requires reasoning about all possible chains of transitions to determine events that might interfere with others.

Parallelism for mobile applications uses operation time and requires sophisticated algorithms since it is not sufficient to run just a few services in parallel. Mobile systems are power constrained but improved wireless connectivity enables shifting computations to servers or the cloud. Leading experts state that, generally, parallel systems can be expected supporting task parallelism and data parallelism, both essential for decentralised and distributed manufacturing applications. Eventually each node of a task can have multiple implementations that target different architecture [4]. For manufacturing applications this allows taking full advantage of the task parallelism on one hand and running independent operations in parallel on the other. Parallelism will revise process planning, for example, by building sequences from independent sub-sequences. For parallelism of operations in manufacturing, industrial networks will strongly rely upon dynamic forms of communication and coordination that handle non-predictable situations by self-adaptiveness and self organization.

4.2.3. Iteration. Developing configuration options and decide about favourable configurations is a highly iterative process and not a straight-line journey. Loops back are possible, as factory and network capabilities identified and may not fit or others may give rise to potential new business opportunities. The 'Iteration' mode emphasises the fact that there is an inherent, evolving nature to structuring. Iteration results in changes that must propagate through the structure's stages, requiring continuous process rework. Within simple settings of collocated operations, the challenge of managing can still be achieved by conventional planning systems and respective intra-organisational decision mechanisms. For networks, management becomes much more complicated, as the involved units and their roles are not stable, but evolve dynamically. However precisely these properties enormously increase a companies' adaptabilities and strongly amplify differentiations and uniqueness. This means continuous restructurings and adaptations for manufacturing networks as well. For the decisions on structuring, re-linking, or breaking up connections in manufacturing networks, iterative procedures

IOP Conf. Series: Materials Science and Engineering **58** (2014) 012002 doi:10.1088/1757-899X/58/1/012002

develop both system structure models and map behaviours onto structures vice versa, ensure the manufacturing networks robustness, their stability against uncertainties, operator mistakes, or imperfections in physical and/or cyber components. Since integration into processes must be orchestrated in order to achieve suitable performance behaviours, it is necessary to ensure the expected alignment with respect to the fit degrees, similar KPI or (estimated values of) key alignment indicators (KAI) [29].

4.2.4. Encapsulation. In general, encapsulation is the inclusion of one thing within another thing so the included thing is not apparent. In DM, encapsulation is concerned with the possible encapsulations of abstractions of units (e.g. models or task descriptions) and transformations (e.g. processes) [17]. The Encapsulation mode enables to build networks and processes by combining elements for creating new processes and units or for atomising units to obtain elements. Self-similarity and compositionality of a unit or a process is a direct consequence of unit- or task encapsulation and provides the basis for constructing networks from components [21] *[xx]*. The models of a unit are accessible through interactions at the interfaces supported by the models. The model element may be seen as based on connectors (links) to construct and compose units. In the tangent space projection, there are two kinds of elements. (I) unit models, and (II) connectors.

The units are loosely coupled and their control is originated and encapsulated by connectors, which is used to define and coordinate the control for a set of components (element or composite). Indeed, the hierarchical nature of the connectors means that composite units are self-similar to their sub-components; this property also provides the basis for hierarchical composition. Each unit model may additionally encapsulate more models and methods.

In a composite, encapsulations in the sub- units are preserved. As a result, encapsulation is propagated in compositions of newly constructed components (units are self-similar) and is also closely related to components' reuse.

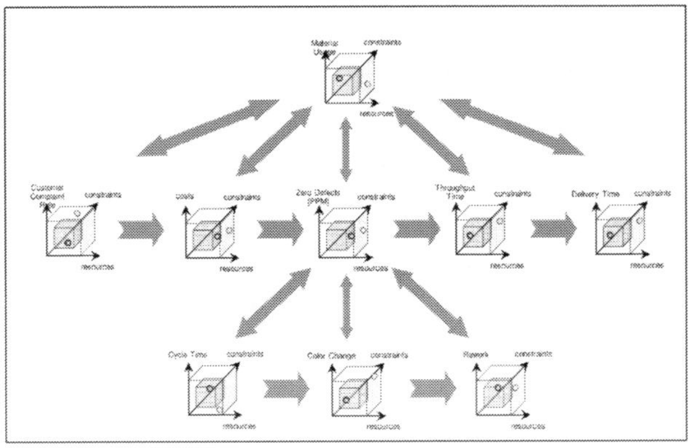

Figure 5. Breakdown (unfolding) of encapsulated behaviour into desired Levels of Detail.

Encapsulated models of units and connectors, may arbitrarily be compressed/broken down resp. fold/unfold (Figure 5). For instance a critical behaviour of a unit on a lower level may have to be compensated on a more aggregated network level or even at the configuration level of the total manufacturing network. Arising criticalities are to be negotiated and harmonized with other units' objectives and resources. A unit's behaviour may generally result in decisions on maintaining the self-organization mode, reducing or removing the autonomy and calling for network interference along the subsequent decision cycle.

Strategy and Objectives.

The network gets vision, mission and network draft which will later be detailed in order to generate the design and the operation parameters. The network strategy has to support the idea that in order to truly align the structure with business requirements, units must be free to negotiate and to choose the solutions that best meet their unique needs.

Monitoring and Analysis.

This stage tracks the execution of the manufacturing processes. It executes by detecting/ sensing the current state of the business and operational manufacturing environment, by monitoring the manufacturing-related business processes for determining if the manufacturing units' behaviours are acceptable (e.g., concerning economic performance), for capturing (unexpected) events and continuously informing on the current situation (e.g., desired, undesired and unexpected events). Activities that constantly update the units' potentials, capabilities or availabilities or that check the network for underperforming units and that notify the network in cases of outages or other alarms, recognised by units' criticalities. Structures, mechanisms and outputs are studied, compared and rated. These analyses may be driven down to sub or sub-sub levels where resource configurations and their contributions to the objectives as well as the SoAs structures (incl. the criticality settings) are broken down. In cases of less severe criticalities, improvements or objectives' alignments are initiated. Severe criticalities will provoke networks' adaptations or reconfigurations.

Network design.

The network is be configured to meet customer requirements best. Partners, units and other actors are identified and linked to a network structure. Processes have to be linked and assigned to responsibilities.

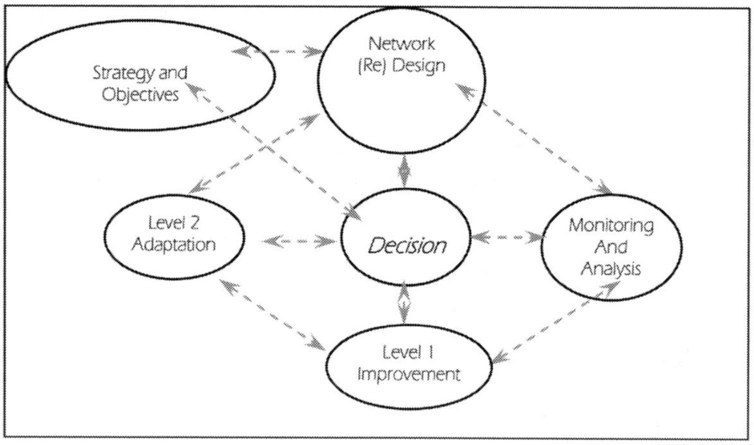

Figure 6. Revolving decision cycle procedure of levelled interventions in manufacturing for gradual continuous configuration.

The strategy elements may be broken down to the decisive factors and the respective indicators that cover all key areas of the networks. They may result in relations of sub objectives and/or aggregated objectives' systems.

Decision.

The decision phase marks the point where the necessary initiatives are taken in order to support the networks evolution into the intended direction. All decisions of importance may be taken, revised, improved or repeatedly cancelled within this cyclic procedure (Fig. 6) i.e. previous program strategy, network configuration, make/buy decision, site decision, process/technology/equipment decisions, etc. are revisited regularly. History and time (complexity attributes) might hinder to execute the resulting decisions immediately. Structures might exist that cannot be instantly eliminated or the building of new competencies will take some time. For the modelling of the network it is therefore recommended to maintain other models (structure simulator) beside the model of the given actual network. These models should provide for "what if" evaluations and simulated comparisons of indicators that make visible, to what extend the actual configuration shows "suboptimal" effects on the results.

Figure 7 illustrates the self-similarity of composite components in a decision network involving the decision cycle as described. Most importantly, every composite component is similar to all sub-components. This means that composition is done in a hierarchical manner. Furthermore, each composition preserves encapsulation. The topological nature ensures that the hierarchical structure of the encapsulation enforces additional rules to ensure the overall process optimum.

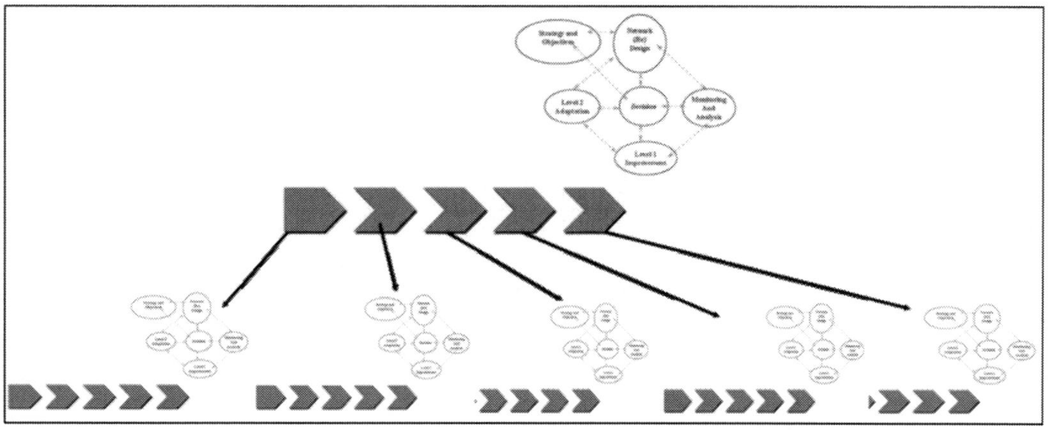

Figure 7. Meshed decision cycles including encapsulated models and instruments to negotiate and decide on manufacturing networks' process fulfilment on several levels of detail according to distributed manufacturing/properties.

A unit component encapsulates all necessary models and procedures. A composite component also encapsulates computation and control [21]. For decentralized decision making based on network business models special logics, algorithms and methods for integration and management seem to be necessary. This concerns the matching of partners as well as the temporary collocation of operations in manufacturing networks. On this basis, all units' behaviour as well as all interrelations may be optimised and planning procedures and logic for the meshed control of configurations, containing processes and resources in networked manufacturing structures, may be established.

4.2.5. Emergence. Emergence focuses on the arising of new patterns, structures and characteristics of networks that are neither really predictable nor fully deductible from antecedent states, events or conditions. DM configurations are ideally envisioned as emergent. Generally, emerging set-ups are characterised as dynamical, meaning they arise over time, as coherent, meaning show somehow enduring integration and occasionally as ostensive, meaning they appear during a set up evolves. In the smart world as outlined, manufacturing processes may therefore be seen as emergent items as well, corresponding to the term emergence precisely in this sense. Complexity science has means to express links and dynamics of interconnectivity, or what in complexity discourse is traditionally termed "emergence"; arising of unforeseen new structures with unexpected new properties e.g.[11].

Example.

In automotive car body shop the model as well as the decision structure was implemented in order to optimize a buffer system and to make the shop more flexible. The objective is to optimise the processes and outcome within a storage area by using adequate commissioning strategies. All objects regularly interact (Iteration) and can simultaneously (Parallelism) be addressed. Every part, every box and also the equipment and the human agents are represented as a Space of Activity (SoA) (Behaviour) and have possible process models available (Encapsulation).

The storage area operates on the basis of commissioning parts. Availability parameters as well as performance indicators are monitored and analysed. Criticalities concerning the spaces of activity activate the decision cycle. The revolving logic switches to the improvement step trying out three options 1 Commissioning of parts, 2 Commissioning of boxes containing respective parts, 3 Manual commissioning depending upon their feasibility. For the cases if commissioning of boxes is sufficient, the system operates on this mode, as long the indicators for the parts commissioning are not more favourable. If the commissioning of boxes runs out of feasibility, it is switch to manual commissioning. The last resort of restructuring would mean a long term investments decision which is done on a more aggregated level. If all of these options fail, the decisions on the supply of material are given to a higher order decision instance, having in either total process/manufacturing network.

The decision cycle will firstly assumes the commission to box as to be supplied to the area, as this is not always possible the option offered different boxes containing their respective parts for picking is chosen, leaving the commissioning to the assembler if this is not possible for as improvement the commissioning is to be manually done by the workers.

The total setup represents a network of mechatronic systems with meshed controlled elements. The above features of the system appear as encapsulating these coupled elements and the networked system. Information processing uses different terminology for control levels. Higher levels are in charge, as soon as the regular process, its improvements and additional efforts are unsuccessful to meet the objectives. The cyclic nature of the decision process was not fully implemented in the practical example (automotive supplier), as the procedure contradicts the time sliced linear overall planning logics in the network (ERP logic) on the more aggregated system levels.

In Figure 7, the process chain emerges as a result of the interactions between units. There is no ultimate configuration solution to be found beyond continuous adaptation and restructuring. To state that process chains emerge, however, does not mean to abandon overall planning. Rather than deriving outcomes by rigid adherence to preconceived strategies, the key for ensuring good solutions is to focus on creating effective rules for interactions. These rules ensure alignments among participants that increase the likelihood of favourable emergent network configuration leading to the objectives fulfilments aimed at.

5. Conclusions and Outlook

What definitely follows for industrial practice is that non-hierarchical views of manufacturing will fully establish. Iterative concepts of planning and control will replace central, sequential, rhythmic and time sliced procedures by event-driven parallel distributed evolving logics. Manufacturing will introduce and apply new types of methods and tools, supporting linkage and reconfiguration as well as high level plug & produce, plug & participate and concurrent work skills.

Any decisions on implementation of smart units in manufacturing as well as virtual resources as services are elements of a company's core strategy and cannot be delegated to IT experts or service providers. The services required by manufacturing differ from general services. The main points, which highlight manufacturing, are interaction ability; powerful functionality (manufacturers will want to streamline business processes and to optimize inventory), real-time ability and Multi Corporation set up.

More standards on all levels will be defined, most likely on international level and done by institutions outside of manufacturing. For implementation decisions, it's rather a matter of choosing and evaluating than developing own standards or engaging in standardisation organisations. It is always worthwhile to keep an eye on rapidly spreading devices of telecommunication and respective freeware for general use that could eventually establish irresistible quasi- or de facto standards.

Manufacturing near associations ought to provide recommendations which existing or upcoming standards should be considered.

Many virtualization instruments address manufacturing main processes, hence key productivity issues. The use of resources along these models will translate into lower costs for all involved units. Early adopters of such novel distributed manufacturing options might immediately set so far unseen KPI benchmarks and cause competition pressure. Specialization of manufacturers using complex and expensive machinery or factories to develop certain products or sub-products for other manufacturers is facilitated.

Moreover these systems might instantly demonstrate drastic changes in the forms of manufacturing or manufactured products and, especially, could initiate novel business models synthesising new services and new products. Business model development should focus on the research questions [37]. Why would those involved in this business model choose a smart operating environment over a traditional manufacturing environment? How will equity be assured when value is delivered as a result of shared-interest, multiple-party work? How should IP be handled in collaborative environments?

Especially, Cloud Manufacturing allows easy integration of applications and processes both within an organization and between different organizations that wish to collaborate. However, some of the greatest concerns are security problems, loss of control (infrastructure, services, and management), technology, difficulty in migrating to other platforms, and loss of reliability. Companies may feel most attracted to the hybrid cloud, an option that might be reserved for applications, which do not require any synchronization or highly specialized or expensive equipment. Initially, hybrid solutions with large portions of proper company implementations are expected.

The development will challenge HR policies. If comparisons to mechatronics and distributed manufacturing hold, there will be mainly high/ICT skilled experts around in these virtualized manufacturing areas. General use of smart services should be sensitive to potential cultural and organizational differences in users' motivation to participate. Collaboration is not always considered appropriate and reasonable across cultures. Control beliefs should be encountered by informing and enabling users. An abort function or a similar type of "emergency feature" to cut off the provider from the user's smart object can easily be installed. Another major finding is that companies should train and provide the entire staffs, not just the "front-line employees, in order to improve their social interaction skills in this technology-mediated service setting. It is important to understand that the introduction of smart interactive services substantially changes the way providers, manufacturers and customers interact [39].

Apart from the possible criticism for its novelty, there is an open controversy with this kind of manufacturing virtualization and all similar approaches, in which IT has a starring role. However, the actual virtualization wave in manufacturing is, despite of obvious gaps and high risks, unstoppable, because the advantages of these technologies are evident and their benefits are indispensable for manufacturing. Mechatronics and distributed manufacturing are intensively working with intelligent devices and units that are networked, so these fields have provided substantial work to incorporate network principles in processes and process control. Advanced implementations from these fields may be seen as the manufacturing of tomorrow in the nutshell where managements don't take but rather game their decisions. Generalising important chapters of this work could bring multiple advantages when consequently expanded into the manufacturing world in total.

References

[1] Adelstein F, Gupta S K S, Richard G G, Schwiebert L 2005 *Fundamentals of Mobile and Pervasive Computing* (McGraw-Hill)

[2] Bongard J 2009 Biologically Inspired Computing *IEEE Computer* **April 2009**, pp 95-98

[3] Buyya R, Yeo C S and Venugopal, S 2008 Market oriented cloud computing, vision, hype, and reality for delivering it services as computing utilities *CoRR*.

[4] Catanzaro B, Fox A, Keutzer K, Patterson D, Yiing S B, Snir M, Olukotun K, Hanrahan P and

Chafi H 2011, Ubiquitous Parallel Computing from Berkeley, Illinois and Stanford, *White Paper*

[5] CERP-IoT 2009, Internet of Things Strategic Research Roadmap 15 SEPTEMBER, *CERP-IoT* (SRA Information Desk European Commission - Information Society and Media DG)

[6] Chan C, Chow H, Siu W, Hung L N, Chu T, Chan H 2012, A Multi-Agent-based RFID Framework for Smart-object Applications *IMECS 2012* (Hongkong)

[7] Derler P, Lee E A, Sangiovanni-Vincentelli, A. 2012 Modeling Cyber-Physical Systems, *Proceedings of the IEEE (special issue on CPS, Ptolemy Project)*, **100(1)**13-28, January 2012

[8] Eguchi A, Thompson C. 2011 Smart Objects In A Virtual World *International Journal of Computer Information Systems and Industrial Management Applications* ISSN 2150-7988 Vol **3** (2011)

[9] EPICS 2013, http.//www.aps.anl.gov/epics/, last retrieved October 15th 2013

[10] Ferguson G T 2002 Have Your Objects Call My Objects *Harvard Business Review* **80 (6)** 138–144

[11] Goldstein J 1999 Emergence as a Construct. History and Issues, Emergence. *Complexity and Organization* **1 (1)** pp49–72

[12] IEEE 1990 *Standard Computer Dictionary* - A Compilation of IEEE Standard Computer Glossaries IEEE (Institute of Electrical and Electronics Engineers)

[13] ISO 1999, ISO 14258, *Industrial automation systems and integration – Concepts and rules for enterprise models*, ISO TC 184 SC5

[14] Kortuem G, Kawsar F, Fitton D and Sundramoorthy V 2010 Smart Objects as Building Blocks for the Internet of Things, *IEEE Internet Computing* Vol **14 no. 1**, Jan./Feb. 2010, pp. 44-51

[15] Kosanke K 2006 *Interoperability of Enterprise Software and Applications* Eds Konstantas et al. (Berlin Springer 2006)

[16] Kuehnle H 2006 A model system approach to distributed manufacturing *Proceedings of CamSIM 2006 Managing Global Manufacturing and Supply Networks* (Cambridge)

[17] Kuehnle, H. 2010 *Distributed Manufacturing – Paradigms, Concepts, Solutions and Examples* (Springer London)

[18] Kuehnle H 2012 Towards Production Network (PN) Theory. Contributions from Systems of Models, Concurrent Enterprising and Distributed Manufacturing *International Journal of E-Business Development* **2(2)** pp 53–61

[19] Kuehnle H, Dekkers R 2012 Some thoughts on Interdisciplinarity in Collaborative Networks' Research and Manufacturing Sciences, *Journal of Manufacturing Technology Management, Emerald* Vol **23**, 2012, **8**, pp 961-975

[20] Langheinrich M, Mattern F, Römer K and Vogt H 2000 First Steps Towards an Event-Based Infrastructure for Smart Things *Ubiquitous Computing Workshop (PACT 2000)* (Philadelphia)

[21] Lau K K, Faris M 2007 Data Encapsulation in Software Components, ed H W Schmidt et al. *CBSE 2007, LNCS 4608* pp 1–16 (Springer Berlin)

[22] Lee E A 2008 Cyber physical systems. Design challenges *ISORC '08, Proceedings of the 11th IEEE Symposium on Object Oriented Real-Time Distributed Computing* IEEE Computer Society (Washington DC) pp 363–369

[23] Macia-Perez F, Berna-Martinez J V, Marcos-Jorquera D, Lorenzo-Fonseca I and Ferrandiz-Colmeiro A. 2012 A New Paradigm. Cloud Agile Manufacturing *International Journal of Advanced Science and Technology* Vol **45**, August 2012 p 47

[24] Meier M, Seidelmann and J, Mezgár I. 2010, ManuCloud. The next-generation manufacturing as a service environment, *European Research Consortium for Informatics and Mathematics News, 2010*

[25] Ming Z, Chunyang H 2013 Research and realization of Resource Cloud Encapsulation in Cloud Manufacturing *IJCSI International Journal of Computer Science Issues* Vol **10** Issue 1, No

3, January 2013

[26] NIST 2011, NIST cloud computing reference architecture, *Special Publication 500-292*

[27] OASIS reports, http.//docs.oasis-open.org/wsrf/wsrf-ws_resource-1.2-spec-os.pdf last retrieved October 15th 2013

[28] Open China ICT 2013 *Document Number D2.1-European Roadmap Survey* draft (European Commission)

[29] Piedade R F, Azevedo A and Almeida A 2012 Alignment prediction in collaborative networks, *Journal of Manufacturing Technology Management* Vol **23** Iss. 8 pp1038 - 1056

[30] Ptolemy 2013, http.//CyberPhysicalSystems.org/CPSConceptMap.xml last retrieved October 15th 2013

[31] Satyanarayanan M 2002 A Catalyst for Mobile and Ubiquitous Computing *IEEE Pervasive Computing* 1 (1) pp 2–5

[32] Sztipanovits J 2007 Composition of cyber-physical systems *Engineering of Computer-Based Systems, ECBS '07, 14th Annual IEEE International Conference and Workshops* pp 3 –6

[33] Sztipanovits J, Koutsoukos X, Karsai G, Kottenstette N, Antsaklis P., Gupta V, Goodwine B, Baras J and Wang S 2012 Toward a Science of Cyber–Physical System Integration, *Proceedings of the IEEE* Vol **100** IEEE No. 1

[34] Ueda K, Hatono I, Fujii N and Vaario J 2000 Reinforcement learning approaches to biological manufacturing systems *CIRP Annals – Manufacturing Technology* Vol **49** No 1 2000 pp 343-346

[35] VDI/VDE 2013 Thesen und Handlungsfelder, Cyber-Physical Systems. Chancen und Nutzen aus Sicht der Automation (Verein Deutscher Ingenieure e.V. VDI/VDE-Gesellschaft Mess- und Automatisierungstechnik (GMA) Düsseldorf) April 2013

[36] Wolf W 2009 Cyber-physical systems *Computer* **42(3)** pp 88 –89

[37] Wu D, Greer M-J, Rosen D W and Schaefer D 2013 Review Cloud manufacturing. Strategic vision and state-of-the-art *Journal of Manufacturing Systems (in press)*

[38] Wu D, Lane J T, Rosen D W and Schaefer D 2012 Towards a cloud-based design and manufacturing paradigm. Looking backward, looking forward, *Proceedings of the ASME 2012 International Design Engineering Technical Conference & Computers and Information in Engineering Conference IDETC/CIE 2012 August 12-15* 2012 (Chicago)

[39] Wuenderlich N-V, v. Wangenheim F and Bitner M J 2011 High Tech and High Touch. A Framework for Understanding User Attitudes and Behaviors Related to Smart Interactive Services *Journal of Service Research* (SAGE Publications)

[40] Yang H C 2010 Cloud manufacturing is a manufacturing service *China Manufacturing Information* No **2** 2010 pp 22-23

[41] Zhang L, Luo Y, Tao F, Ren L and Guo H 2010 Key technologies for the construction of manufacturing cloud *Computer Integrated Manufacturing Systems* Vol **16** No **11** 2010 pp 2510-2520

Effect of Addition Pb Content on Lead Titanate Thin Films for MIM Capacitor

Z. Nurbaya[1,2,*], and M. Rusop[1]

[1]NANO-ElecTronic Centre (NET), Faculty of Electrical Engineering, Universiti Teknologi MARA, 40450 Shah Alam, Selangor, Malaysia

[2]Razak School of Engineering and Advanced Technology, Universiti Teknologi Malaysia, 54100 Kuala Lumpur, Malaysia

*E-mail: nurbayazainal@gmail.com

Abstract. Lead titanate ($PbTiO_3$) has been precisely studied on the films preparation due to its unique perovskite structure having ferroelectric behaviour with high permittivity factor for complex storage applications such as capacitor, transistor, and ferroelectric random access memory. In the need to have high quality ($PbTiO_3$) thin films, the lead (Pb) volatilization during crystallization process by means of applying heat treatment as part of crystal growth purpose must be realized. Therefore addition of Pb excess is the highest consideration in lead based composite to ensure perfect crystal growth ever form as well as yield to great particles polarization. There are some issues like fatigue, imprinting, and leakage current that need to be considered to assess the thin films performance. However, the present study focuses only on the crucial issue of leakage current behaviour of the nanoscaled thickness that might be in relation to the microstructural properties of the thin films. Thus, the influence of having extra Pb content might ascertain the benefits of $PbTiO_3$ thin films towards high performance metal-insulator-metal capacitor.

Keywords: Lead titanate thin films; extra lead content; surface morphology; leakage current density behaviour; nanoscaled thickness films; resistivity behaviour

1. Introduction

Lead based ceramic materials have been discovered since more than 30 years ago. Lead titanate $PbTiO_3$ with the unique perovskite structure promotes broad range of device application due to the outstanding properties; one of them is high dielectric permittivity. Variety techniques of $PbTiO_3$ thin films have been prepared by using radio frequency magnetron sputtering [1]–[3], metal organic chemical vapor deposition, MOCVD [4]–[6], pulsed laser deposition, PLD [7]–[9], and sol-gel derived spin coating method which involves solution to gelatinous films transformation by using spin coating technique [10]–[13]. Sol-gel process offers ability to control the stoichiometry composite, obtain extremely thin films homogeneity, deposition of large area, and can be done at room temperature. These being the reasons to in line with the cost effective and time efficient in producing high quality thin films performance. Therefore, the solutions need to be carefully prepared in order to gain high purity elements prior to films deposition. It is compulsory in sol-gel process to be familiar with the type of starting materials, the range of molar concentration, and also the addition of extra Pb excess (5~25mol%) purposely for preparing $PbTiO_3$ films [2], [14], [15]. The purpose of adding Pb excess is to compensate the Pb deficiency in the thin films during high heat treatment that causing to the defect of A-site, ABO3-perovkite structure which leads to the permanent failure as ferroelectric material.

In this study, the $PbTiO_3$ thin films were prepared by sol-gel derived spin coating technique with extra Pb excess (5~20mol%), to reveal the effect towards nanostructural changed and electrical properties.

2. Experimental Method

This study presents the actual experimental investigation towards characterization of $PbTiO_3$ thin films based on simple configuration of metal-insulator-metal, MIM for capacitor application. Basically, solution deposition method through spin coating techniques is the most preferable technique to produce large deposition area of highly resistive materials as well as keep in pace with industrial needs. Thus, low cost production with zero defect performance could yield to major achievement in the most discrete device fabrication challenges. Here, we followed the simple process of solution-to-gelatinous transformation according to Bao *et al* procedure with alternate starting materials which can be referred in [10].

2.1. PbTiO3 solutions

The $PbTiO_3$ precursor solutions were prepared by simple sol-gel method in air mixed-oxide, at room temperature. The starting materials involved alkoxide material, Titanium tetraisopropoxide, TTiP (Sigma Aldrich, 97%) and lead acetate trihydrate (Sigma Aldrich, 99.9%) with the solvent, 2-methoxyethanol (Sigma Aldrich, 99.9%). First, the lead acetate trihydrate powders were dissolved in 2-methoxyethanol stirred at 100°C for about 1hour. Then, titanium isopropoxide was added to the solutions and stirred continuously for the next 1hour. The solutions were first prepared at different concentration (0.1~ 0.5mol) in previous study as reported elsewhere. It was found that 0.4mol solution is the most preferable concentration that brings to extended investigation that precisely makes for the present study. After that, the prepared solution was added with extra Pb excess (10 and 20mol%) referred to the actual concentration, named the samples as PbEq, Pb10, and Pb20. Then, a little drop of methanol was added before the solutions undergone for a day ageing process at room temperature.

2.2. Thin films depositions and characterization

The deposition process of $PbTiO_3$ was setup under 3000rpm spin speed for about 30s onto cleaned 200nm indium tin oxide (ITO) coated nglass substrate. Prior to the deposition process, the solutions were putted in ultrasonic bath at 50°C about 20mins to form even smaller particulates ensured nice dispersion of multicoated thin films. Then, the prepared films were dried to remove residual organic matters at 150°C for 10mins and continued for five times repeated coating at same conditions. Lastly, in crystallization process, the films were annealed at 500°C for 30mins and let them cooled off at room temperature. These two steps of heat treatment process are to ensure the films were under crack free conditions.

The characterization of $PbTiO_3$ thin films was carried out for electrical and structural investigation with gold (Au) as metal contact based on configuration Au/$PbTiO_3$/ITO/glass substrate illustrated in Fig. 1. The thin films characterizations were specifically investigated by two point probes solar simulator (BUKOH KEIKI CEP-2000) to determine thin films resistivity, surface profiler (Veeco) for film thickness measurements, and atomic force microscopy, AFM (Park System XE-100) and field emission scanning electron microscopy respectively (FESEM-JEOL) for structural of thin films.

3. Result and Discussions

3.1. Structural Properties

Preparation of the thin films at 0.1~0.5 mol is to ensure that the initial concentration can be maintained throughout developing the $PbTiO_3$ thin films for storage application mainly as capacitor discrete device. Such that, the surface morphology of 0.1 to 0.5mol concentration were observed by AFM with 10μm of scan size and 2.50 of scan rate as shown in Figure 1. It is revealed that the average surface roughness can be as low as 0.388nm found at 0.4mol concentration films. At 0.5mol concentration, the films tend to have such high roughness about 5.255nm due to the large gap of hills and valley at surface films. Entire results of average surface roughness at various molar concentrations were tabulated in Table 1.

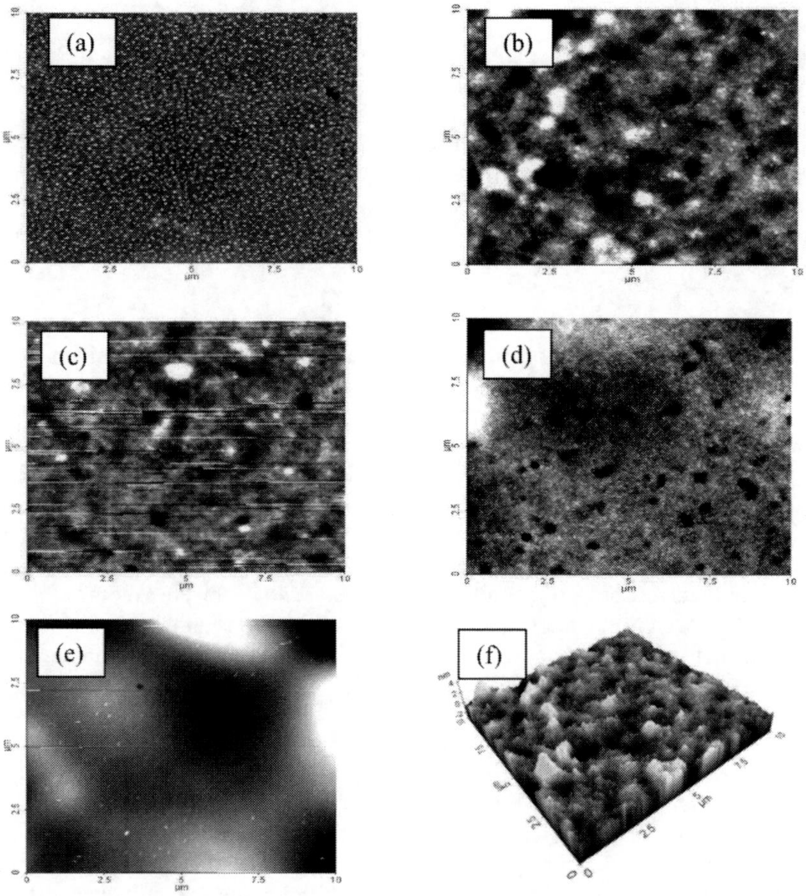

Figure 1. AFM result of $PbTiO_3$ thin films varied at different molar concentration, (a) 0.1mol, (b) 0.2mol, (c) 0.3mol, (d) 0.4mol, (e) 0.5mol, and (f) 3-D view of 0.4mol.

Table 1. Average surface roughness of $PbTiO_3$ thin films at various molar concentrations.

Molar concentration (mol)	Average surface roughness
0.1	1.057 nm
0.2	0.593 nm
0.3	0.482 nm
0.4	0.388 nm
0.5	5.255 nm

Thus, 0.4mol prepared $PbTiO_3$ solutions were then ready for three samples that simply according to the addition of Pb content prior to one day ageing process, known as PbEq, Pb10, and Pb20. Instead of identifying the RMS values of surface roughness, the prepared films were through FESEM observation in order to assess preferred grains size, less grain boundaries existence, also densification of the grains. Thus, Figure 2 shows the surface morphology and cross-sectional view of the films surface.

From the figure, it is clearly seen that the grains growth have much influenced by the used of ITO electrode. The nanostructured $PbTiO_3$ thin films were obtained very high dense films for PbEq without having actual grain structure. On the other hand, seems that both Pb10 and Pb20 samples were having amorphous structure shows the presence of combined flakes and grains distribution on films surface. However, Pb10 sample managed to performs such a uniform structure compared to Pb20 samples where flakes , whereas the Pb10 sample was having uniform gain distribution, though multi structures formation like combination of flakes and grain structure was obtained due to the influenced of ITO surface that has probability of high surface roughness. Therefore, Pb10 sample was carried out for grain size measurement that achieved around 55.65nm that in well agreement with the study worked by Cheng et al; 50nm grain size of polycrystalline films [16]. Moreover, from the cross sectional view of the films in Figure 2(d), Pb10 sample formed a highly densification films along with thickness measurement of 418nm. Entire films thickness measurements are tabulated in Table 2 along with the resistivity and conductivity results.

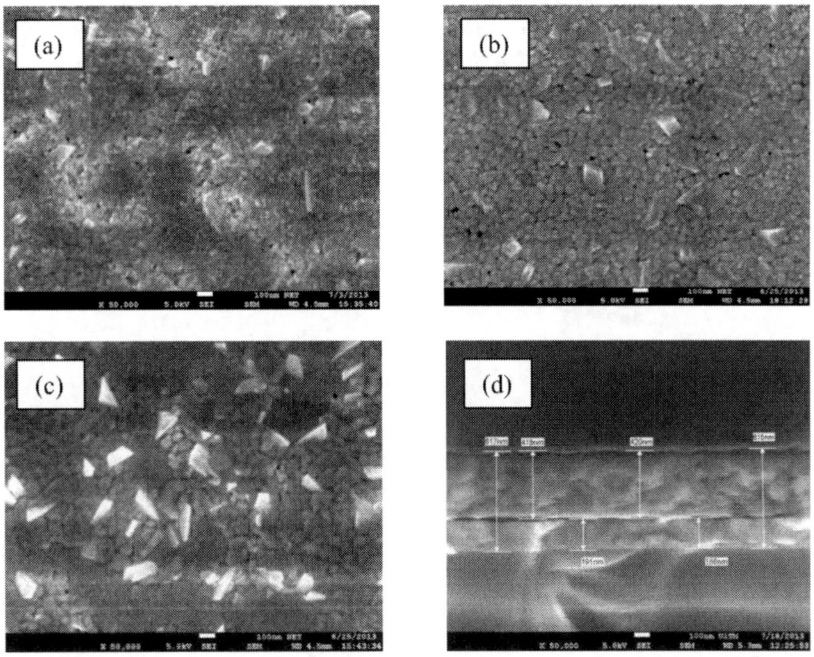

Figure 1. FESEM result of $PbTiO_3$ surface morphology at influence of addition Pb content (a) PbEq, (b) Pb10, (c) Pb20, and (d) cross-sectional view of Pb10. (x50,000 magnification)

3.2. Electrical Properties

3.2.1. Resistivity behaviour. In this investigation, the resistivity and conductivity were calculated and tabulated in Table 2. The reason to have resistivity behaviour of the thin films is to show the interesting factor that employing lead titanate for MIM capacitor in general. Theoretically, this material typically behaves insulate to the passage of heat and electricity at films thickness dependency [17].

Table 2. Summary of Films Thickness, Resistivity, and Conductivity Behaviour of $PbTiO_3$ thin films at Influence of Extra Pb Content.

Sample	**Thickness**	**Resistivity**	**Conductivity**
PbEq	239 nm	2.200×10^5 Ω.cm	4.544×10^{-6} $S.cm^{-1}$
Pb10	418 nm	3.035×10^5 Ω.cm	3.295×10^{-6} $S.cm^{-1}$
Pb20	359 nm	2.731×10^5 Ω.cm	3.660×10^{-6} $S.cm^{-1}$

The result shows that highest films resistivity was obtained at Pb10 samples (3.035×10^5Ω.cm) which consisted to the thickness measurement (418nm). In other words, addition of 10mol% extra Pb content offers high densification addition to less grain boundaries resulted to high resistivity films.

3.2.2. Current density behaviour. In actual I-V characteristic in ferroelectric thin films are known as total switching current (J_S) that performed by polarization current (J_P) and electronic current (J_E) as previously stated in [18]. Besides that, there are other things need to take into account at such as, the true leakage current (J_L) and also the current that may be affected by ageing factor (J_A) as proposed in [19]. Park *et al* in his excellent study on the effect of oxygen vacancy in ferroelectric perovskite materials will permanently leads to fatigue, imprinting, and leakage current one of them [20]. Herein, the issue of leakage current in thin films processing is the most critical concern in memory device application.

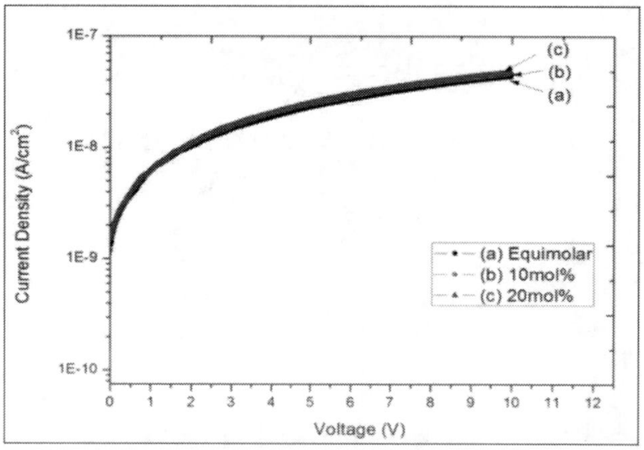

Figure 3. Plot of current density of $PbTiO_3$ thin films measured under 10V

Therefore, this study managed to obtain current-voltage characteristic (log *J* versus *V*) result as depicted in Figure 3 at such low leakage current density found under 1.00×10^{-7}A/cm^2. The voltage is applied with a step of 0.1V for each measurement at room temperature. It is noted from the figure that the current is exponentially increase under low applied voltage (5V), followed by a saturation region at higher voltages. Herein, the J-V characteristic in this study were found typically a form of mixed Schottky and Poole-Frenkel emission that reported as MIM capacitor in [13], [18].

However, such low leakage current existed might be due to the presence of interfacial layer between the films and bottom electrode as proven in Figure 2(d). This layer kind of known as a 'dead layer' creates even small conductivity whenever 0.0V being applied to the device. It also consisted to the study reported by Tang *et al* that worked on the electrical behavior of calcium doped leat titanate films [22]. Apart of that, Chiu *et al* discovered that the leakage current level is high in most polycrystalline structure (~10^{-4}A/cm^2) for Ta_2O_5 thin films [21] in relation with the improvement made by Jia *et al* in [23] that used Ru_2O as bottom electrode for polycrystalline bilayer films which had captured much interest on this study to keep going on having bilayer configuration in future.

4. Conclusion

Briefly, $PbTiO_3$ thin films were successfully prepared through economical and low temperature crystal growth of sol-gel derived spin coating method. This work highlights the significant effect of adding extra Pb content towards resistivity and leakage current density behaviour of the nanoscaled thickness thin films MIM capacitor. This is one of the criteria need to be acknowledged as part of research work related to ferroelectric ceramic material.

Here, Pb10 samples have demonstrated polycrystalline structure as well as denser films observed by FESEM images at x50,000 magnification and found 55.65nm of grain size. This promotes to the electrical properties measured by I-V measurement that the Pb10 films having very high resistivity value; $3.035\text{x}10^5$ Ω.cm at 418nm thickness as well as low leakage current density; $1.00\text{x}10^{-7}$A/cm^2 under ohmic region. Therefore, addition of 10mol% Pb content is the most preferable amount that promotes as great dielectric layer in high MIM capacitor performance.

Acknowledgement
The author would like to thank NANO-SciTech Center (NST) for using AFM, Research Management Institute (RMI) of Universiti Teknologi MARA (UiTM), Ministry of High Education (MOHE) for the financial support.

References

[1] Hwang H S, Park Y and Choi W S 2008 *Microelectron. Eng.* **85**, 2456.
[2] Chang W L and He J L 2004 *J. Electroceramics* **13** 35.
[3] Iijima K, Tomita Y, Takayama R and Ueda I 1986 *J. Appl. Phys.* **60**, 361.
[4] Chen, H Y. Lin, J Tan K L and Feng, Z. C. 1996 *Thin Solid Films* **289** 59.
[5] Sakurai A, Nakaiso T, Nishida K, Ando A and Sakabe Y 2004 *Key Eng. Mater.* **269** 61.
[6] Vellaikal M and Kingon A I 1996 *Thin Solid Films* **287** 139.
[7] Chaoui N, Millon E, Muller J F, Ecker P, Bieck W and Migeon H N 1999 *Mater. Chem. Phys.* **59** 114.
[8] Khan M A, Comyn T P and Bell A J 2008 *J. Eur. Ceram. Soc.* **28** 591.
[9] Pandey S K, James A R, Prakash C, Goel T C and Zimik K 2004 *Mater. Sci. Eng. B* **112** 96.
[10] Bao D, Yao X, Wakiya N, Shinozaki K and Mizutani N 2002 *Mater. Sci. Eng. B*, **94** 1.
[11] Kim C J, Lee W J, and No K 1998 *Thin Solid Films* **312** 1–5.
[12] Tahan D M and Safari A 1992 in *Applications of Ferroelectrics, 1992. ISAF '92* 420.
[13] Yadav H O 2004 *Ceram. Int.* **30** 1493.
[14] Sidorkin A, Nesterenko L, Sidorkin A, Ryabtsev S and Bulavina G 2010 *Solid State Sci.*, **12**, 302.
[15] Yang J, Kim W S and H P U 2000 *Thin Solid Films* **377–378** 739.
[16] Cheng S D, Kam C H, Zhou Y, Que W X, Lam Y L, Chan Y C and. Gan W. S 2000, *Thin Solid Films* **375** 109.
[17] Callister W D 2006 *Material Science and Engineering: An Introduction.*
[18] Chen H, Tsaur S and Lee J Y 1998 *Jpn. J. Appl. Phys.* **37** 4056.
[19] Lee K, Rhee B R and Lee C 2001 *J. Korean Phys. Soc.* **38** 723.
[20] Park C H and Chadi D J 1998 *J. Korean Phys. Soc.*. **32** 143.
[21] Chiu F, Wang J, Lee J Y and Wu S C 1997 *J. Appl. Phys.* **81** 6911.
[22] Tang X G, Wang J, Zhang Y W and Chan H L W 2003 *J. Appl. Phys.* **94** 5163.
[23] Jia Q X, Changb L H and Andersonb W A 1995 *Thin Solid Films* **259** 264.

Examining Users' E-Satisfaction in the Usage of Social Networking Sites; Contribution from Utilitarian and Hedonic Information Systems

Mohd Shoki Md Ariff[1], Tay Kai Shan[1], Norhayati Zakuan[1*], Nawawi Ishak[2], Mohd. Ridzuan Wahi[2]

1 Department of Business Administration, Faculty of Management, Universiti Teknologi Malaysia (UTM), Skudai, Johor, Malaysia
2 Lembaga Tabung Haji Malaysia

E-mail: norhayatimz@utm.my

Abstract. E-satisfaction (eSAT) is an important success factor of online service providers such as social networking sites (SNSs). The utilitarian and hedonic information systems are crucial in determining users' eSAT of SNSs, especially among young users. The utilitarian aspect of an information system is productivity-oriented which aims to enhance the users' task performance, and it is important in measuring eSAT of SNSs. In this study, the original constructs of Perceived Ease of Use (PEOU) and Perceived Usefulness (PU) of TAM of utilitarian information system was first developed in this research framework. The use of SNSs, such as Facebook, is pleasure-oriented, in which self-fulfilling values to the users are important in determining users' satisfaction towards the SNSs. Therefore, Perceived Enjoyment (PE) of hedonic information system is added to the framework. Thus, the research framework of this study includes both utilitarian (PEOU and PU) and hedonic (PE) aspects of information systems to determine Malaysian young users' eSAT in the usage of Facebook, a social networking site. In this framework, the effects of PEOU, PU and PE on eSAT in the usage of Facebook are examined among Facebook's users in the age of 18 – 24 years old. The effects of PEOU on PU and PE are also examined. Online questionnaire survey was employed and a total of 384 sets of questionnaires were gathered from users of Facebook. The results indicated that PEOU has positive effects on PU and PE in the context of Facebook. In addition, PEOU, PU and PE are also found to have positive effects on eSAT. PE of hedonic information system exerted higher effect on eSAT, compared to PEOU and PU of utilitarian information system, highlighting the importance of pleasure orientation in the usage of Facebook of SNSs. Managerial and theoretical implications of the study are discussed in term of measuring and enhancing users' eSAT in the usage of SNSs, particularly Facebook.

Keywords: Technology Acceptance Model; Perceived Enjoyment; E-Satisfaction; Social Networking Sites – Facebook

1.0 Introduction

The increase adoption of social networking sites (SNSs) is drawing much attention from academic and industry researchers to explore users' behavior in this area [1]. According to Zhong *et al.* [2], the advent of SNSs have drastically changed human interaction and millions of people worldwide are now spending much of their leisure time on it. Facebook which has a high popularity and a world population of about 1.11 billion users [3] is unquestionably a great research platform for researchers to study behavior and satisfaction of users. In Malaysia, Facebook is the top SNS with approximately 13 351 560 monthly active users and the largest age group using it is young users who are in the age of 18 to 24 years old - 4 607 560 users or equivalent to about 34.51% of total monthly active users [4]. However, although many researchers [5, 6, 7 and 8] had studied on the determinant factors of eSAT, there is none in the context of users of Facebook aged between 18 to 24 years old in Malaysia. Hence, more researches examining actual behavior, attitude and satisfaction of this group of users towards SNSs (Facebook) are needed.

Theoretically, the studies of information system have identified two types of information systems; utilitarian information system and hedonic information system which influence users' satisfaction. Previously, many researchers had applied Technology Acceptance Model (TAM) in investigating users' acceptance of utilitarian information system rather than hedonic information system as TAM was first developed to measure utilitarian information system. Nevertheless, due to the evolution of information system settings, many researchers had shifted their attentions from utilitarian information system to hedonic information system [9]. Although social networking site is one of the most popular hedonic information systems, there are relatively limited studies that have applied PEOU and PU of TAM in the context of SNSs. Many researchers [10 and 11] had found that PEOU increases PU in different information systems, and this led to the users' satisfaction. Sibona and Choi [7] highlighted that the same positive effects do exist in the usage of Facebook in United States, and PEOU of SNSs was found to have significant positive effect on PU in the context of university students in Malaysia [12]. On the other hand, in order to increase the validity of research framework in the context of users' satisfaction of SNSs, it is ought to include both utilitarian aspects (PEOU and PU) as well as hedonic aspects, such as perceived enjoyment (PE). Therefore, PE has been added to form hedonic information system in examining user satisfaction of SNSs. Previous researchers [13 and 14] highlighted that PEOU has positive effect on PE in various research backgrounds but relatively less in the context of SNSs. Furthermore, many studies [14, 15 and 16] have been focused on the impacts of TAM and PE on users' behavioural intentions rather than their actual behaviour such as e-satisfaction (eSAT), in which, according to Taghizadeh *et al.* [8], is one of the major component used to evaluate the success of an information system. In Malaysia, there is less research studying on the relationship between the two variables (PEOU and PU) in the context of Facebook users within 18 to 24 years old, either students or non-students. Most of the studies did not specify Facebook as the only research area but apply for all types of social networking sites available in the nation. For instance, Leng *et al.* [35] had found that perceived ease of use has significant positive effect on perceived usefulness of SNSs in the context of 283 university students in Malaysia. Hence, a study must be carried out to fill this gap by examining the effects of TAM and PE on eSAT. Therefore, a research is needed to determine the interrelationship between PEOU and PU of utilitarian information systems and PE of hedonic information system, as well as their impact on user satisfaction in the usage of SNSs (Facebook) among Malaysian young users. Thus, this study enriches the use of utilitarian and hedonic aspects in the usage of SNSs and in examining their effects on users e-SAT in Malaysia context, specifically among young users.

2.0 Literature Review

Kwon and Wen [17] defined SNS as a web-based service created based on certain meaningful and valuable relationships such as friendship and common interests. Generally, a SNS can be very useful for individual or organizations if it is used properly. According to King and He [18], the acceptance and adoption of information systems have been studied by many researchers in various contexts such as work-related systems, general systems, SNSs and also e-commerce. Two types of information systems – utilitarian and hedonic – have been identified as major components influencing user attitude and actual behavior, such as e-satisfaction. According to van der Heijden [13], utilitarian information systems are productivity-oriented while hedonic information systems are pleasure-oriented. In other words, utilitarian information system and hedonic information system are systems with different constructs in which the former focuses on productive use while the latter focuses on the aspect of fun and enjoyment. Previously, many researchers were focusing their studies on utilitarian information systems. However, due to the emerging of hedonic information system in recent years, many researchers have shifted their attentions and interest from utilitarian information system to hedonic information system [9]. Nevertheless, most of the theories, such as Technology Acceptance Model (TAM) were originally developed to measure the technology acceptance of utilitarian information system rather than hedonic information systems [19]. Hence, there are arguments on whether the same technology acceptance theories are applicable for both utilitarian and hedonic information systems.

SNS is considered as hedonic in nature and therefore, it was unsuitable to apply utilitarian-based research model in the context of SNSs [20]. Although SNS is categorized under pleasure-oriented information system, it has consistently enhanced its features and functions so as to increase users' adoption on it. For instance, Facebook is capable of sharing a variety of information by updating status, blogs, images, videos and music [21]. Furthermore, SNSs also offer a great communication platform for all the Internet users. Therefore, SNSs are actually more than just a pleasure-oriented information system. According to Ernst *et al.* [19], Facebook is a combination of hedonic and utilitarian information system because it is able to create fun to its users and at the same time allows its users to send files, attach files and even discussion in a group. Hence, it is more appropriate for researchers to include both hedonic and utilitarian aspects when examining acceptance and satisfaction of users towards the sites.

TAM is an information systems acceptance theory which was developed by Fred Davis in 1985 to explain computer usage and the acceptance of information technology. Basically, TAM consists of two technology acceptance measures: perceived ease of use (PEOU) and perceived usefulness (PU). PEOU refers to the degree to which a user believes that using a particular system will be effortless while PU is defined as the users' perception to the extent that the use of a particular system will improve his or her job performance [22]. PEOU has been regarded as a factor that influences users' satisfaction, and it focuses on avoiding irritation in order to enable users to understand the website layout and navigation in their search for products and services that are offered at the website [37]. As PU involves users' judgment on the extent to which using an information system improve their job performance, therefore, designing a simple website with a clear structure and developing a credible fulfillment are essential [38]. Apparently, PEOU and PU were first developed to measure the acceptance of utilitarian information systems rather than hedonic information systems. Nevertheless, according to Koch et al. [23], TAM does include "external variables" that suit to any features of a particular system. Similarly, Venkatesh and Davis [24] proposed that there is a need to extend TAM to include other variables such as subjective norm, extrinsic motivations, enjoyment and et cetera in order to increase the validation of TAM in different types of information systems.

According to Oliver [29], satisfaction can be defined as the "the summary psychological state resulting when the emotion surrounding disconfirmed expectation is coupled with the consumer's prior feelings about the consumption experience." It implies that customers are satisfied if their consumption experience meets their needs and expectations [30]. In the electronic mediated environment, the extent to which users believes that an information system is capable of fulfilling their information requirements influence their e-SAT. Therefore, e-SAT is referred to customer satisfaction in the online market-space, influenced by customers' experience and knowledge in the usage of an information system; it is about consumers' judgment of their Internet retail experience [36].

Although e-satisfaction is vital in determining the success of an information system, there are relatively less researches on the factors influencing the eSAT in social network setting. Nevertheless, most of the previous studies [25, 26, and 27] were focused on the behavioral intentions of users in social network adoption and acceptance through TAM and PE. Hence, this study attempts to fill this gap by investigating the effects of TAM and PE on eSAT. Table 1 shows the contribution of TAM and PE on the acceptance and users' eSAT in the context of SNSs.

Other than the external variables, the final version of TAM also indicates that the original constructs of TAM: PEOU and PU do have relationships among themselves, and most researches indicated that PEOU positively affect PU in the usage of information systems [7, 10, 11, 12, 25 and 26]. Thus, a hypothesis is established: H_1: PEOU has positive effect on PU of Facebook among Malaysian young users.

The main goal of hedonic information system (e.g. SNS) is to provide fun and enjoyment to its users [27]. Perceived enjoyment (PE) has been widely used as an important determinant of the acceptance and satisfaction of hedonic information systems in previous studies. PE refers to the extent to which the activity of using a computer system is perceived to be personally enjoyable in its own right aside from the instrumental value of the technology [24]. Furthermore, PE was also introduced by Venkatesh and Davis [24] as an external variable of TAM. Since hedonic information system is primarily associated with fun and pleasure, PE would be an appropriate external variable in measuring the acceptance and satisfaction of users. A number of previous studies had been focused on investigating the effect of PEOU on PE. According to Gu *et al.* [28], users who perceive that using a system is effortless will positively affect their PE. Moqbel [27] stated that the main goal of users to use SNSs is to enjoy and have fun while PEOU is an important element that helps them to attain the goal. Therefore, H2 is proposed: PEOU has positive effect on PE of Facebook among Malaysian young users.

Table 1: Contribution of PEOU and PU of TAM (Utilitarian information system) and PE (Hedonic information system) to the Acceptance and E-Satisfaction (eSAT) of Social Networking Sites (SNSs)

Previous Studies	**Variables**		
	PEOU	**PU**	**PE**
Kim *et al.* (2010)	√	√	√
Ong and Day (2010)			√
Curran and Lennon (2011)	√	√	√
Nistor (2011)	√	√	
Nithinant *et al.* (2011)	√	√	
Gómez-Borja *et al.* (2012)	√	√	
Gong (2012)			√
Lane and Coleman (2012)	√	√	
Murad Moqbel (2012)	√	√	√
Sibona and Choi (2012)	√	√	
Ernst *et al.* (2013)	√	√	√

Note: PEOU-Perceived Ease of Use; PU-Perceived Usefulness; PE-Perceived Enjoyment

According to Sibona and Choi [7], the Facebook users are satisfied with the site when they find it easy to use and useful, and PEOU may lead to higher eSAT. Hence, the sub hypotheses are proposed: H_{3a}: Perceived ease of use has positive effect on the users' e-satisfaction of Facebook and H_{3b}: Perceived usefulness has positive effect on the users' e- satisfaction of Facebook. Similarly, PE is also applied in numerical studies to investigate its impact to the adoption and intention to use a system. Most of the studies [14, 15, and 16] indicated that PE is positively related to the users' intention to use a system. In addition, some researchers [5 and 6] have shown that PE is a key element in predicting users' eSAT in the usage of a particular system. Apart from these, previous studies on SNSs [13, 15, 25 and 27] highlighted the positive effect of PE on the adoption of SNSs. With these, the researchers assume another sub hypothesis of H_{3c}: Perceived enjoyment has positive effect on the users' e-satisfaction of Facebook. Lastly, the three sub hypotheses are summarized and a general hypothesis is proposed for the purpose of predicting the effects of PEOU and PU of TAM and PE on users' eSAT in the context of Facebook: H_3: PEOU and PU of TAM and PE have positive effects on the users' e-satisfaction of Facebook among Malaysian young users.

3.0 Methodology

Review of the literature indicated that a research is needed to determine the interrelationship between PEOU and PU of utilitarian information systems and PE of hedonic information system, as well as their impact on user satisfaction in the usage of SNSs (Facebook) among Malaysian young users. A questionnaire was developed to measure the utilitarian and hedonic aspects of the SNS. The utilitarian component covers:

- PEOU based on the work of Davis [22] and Liao *et al.* [11]. There are six items developed for PEOU. The items are:
  - I find it easy to learn how to use Facebook.
  - I find it easy to get Facebook to do what I want it to do.
  - My interaction with Facebook is clear and understandable.
  - I find Facebook to be flexible to interact with.
  - It is easy to remember how to use Facebook.
  - Overall, I find Facebook to be easy to use.

- PU with six items. The items developed were adopted from Davis [22], Nithinant *et al.* [31] and Gómez-Borja *et al.* [32]. The six items are:
  - I find the functions of Facebook to be useful for me.
  - I find Facebook to be a useful site for communication.
  - I find using Facebook contributes to interaction with other people.
  - I find using Facebook enables me to access a lot of information.
  - I find using Facebook improves the efficiency in sharing information with others.
  - Overall, I find Facebook to be useful.

PE was used to measure the hedonic aspect of the SNS. It covers five items of PE based on the work of Liao *et al.* [11]. The items are:

- I find using Facebook to be interesting.
- I find using Facebook to be enjoyable.
- I find using Facebook is a good way to spend my leisure time.
- A variety of applications in Facebook arouse my curiosity.
- The process of using Facebook is pleasant.

For E-Satisfaction of users of Facebook, five items used were adopted from Ribbink *et al.* [33] and Shiau and Luo [5]. The items are:

- I am happy with Facebook.
- I am satisfied with the services provided by Facebook.
- I am satisfied with my decision to sign up a Facebook account.
- My choice to sign up a Facebook account was a wise one.

- I think I did the right thing by signing up to Facebook.

The scale of all items of PEOU, PU, PE and e-SAT were measured based on five points Likert ranging from 1-strongly disagree to 5-strongly agree.

The population of this study was all Malaysian Facebook users aged between 18 to 24 years old. With the population size of about 4 607 560 Facebook users [4], the sample size was set to be 384 Facebook users based on the sample size table recommended by Krejcie and Morgan [34]. Convenience sampling was utilized to draw the sample from population due to difficulties to apply probability sampling procedure as well as time and cost constraints. The researchers employed web-based survey by creating the online questionnaires using Google Form at https://docs.google.com/forms/d/1N5ERebWa7bjIeFvXqCRjcG6Qommx1CvAk7INv0qoqhk/viewform and the link of the questionnaires was shared and distributed in Facebook in order to make sure all the respondents were Facebook users. Furthermore, filter questions on the age and nationality of the respondents were included before the respondents could proceed to the survey in order to make sure the questionnaires reached the correct respondents. The data collected was automatically stored in Google spreadsheet and finally exported by the researchers to Statistical Package for Social Science (SPSS) for further analysis.

The Kaiser-Meyer-Olkin (KMO) measures of sampling for PEOU, PU and PE, and e-SAT were 0.948 and 0.902 respectively. Since all KMO values were above 0.70 with Bartlett's test of Sphericity of 0.000, the researchers were allowed to proceed with factor analysis [39]. The Total Variance Explained of Confirmatory Factor Analysis (CFA) indicated that there are three dimensions (PEOU, PU and PE) with eigenvalues greater than 1 have been extracted with the cumulative percentage of 79.741%. The rotated component matrix had yielded three components and all the items in each component were retained since all the factor loadings were above 0.5. For eSAT, there was only one dimension with eigenvalues greater than 1 being extracted in CFA with cumulative percentage of 80.339%. All items of eSAT were accepted with factor loadings of more than 0.5.

In order to measure the degree of consistency and correlations among the items, reliability test was performed. In this study, the alpha values for PEOU, PU, PE and eSAT were 0.942, 0.939, 0.948 and 0.938 respectively. All scores are above >0.7, thus it can be concluded that constructs of PEOU, PU, PE and e-SAT were reliable for further analysis.

4.0 Result and Analysis

The result of regression analysis on the effect of PEOU on PU is presented in Table 2. It was found that PEOU has positive effect on PU (β 0.599, t 14.637, Sig. 0.000). Therefore, H_1 is accepted. On the other hand, the R^2 value shows that 35.9% of the variation in PU is explained by the variation in PEOU.

Table 2: Results on the effect PEOU on PU

Coefficients[a]

Model		Unstandardized Coefficients		Standardized Coefficients	t	Sig.	Collinearity Statistics	
		B	Std. Error	Beta			Tolerance	VIF
1	(Constant)	.964	.158		6.083	.000		
	PEOU	.636	.043	.599	14.637	.000	1.000	1.000
F Statistic = 214.250 R Square (R^2) = 0.359								
a. Dependent Variable: PU								

As shown in Table 3, the regression result indicated that PEOU has positive effect on PE (β 0.564, t 13.362, Sig. 0.000). Therefore, H_2 is accepted. The R_2 value of 0.319 indicated that 31.9% of the variation of PU is explained by the variation in PEOU.

Table 3: Results of the effect of PEOU on PE

Coefficients[a]

Model		Unstandardized Coefficients		Standardized Coefficients	t	Sig.	Collinearity Statistics	
		B	Std. Error	Beta			Tolerance	VIF
1	(Constant)	.773	.194		3.988	.000		
	PEOU	.710	.053	.564	13.362	.000	1.000	1.000
F Statistic = 178.541								
R Square (R^2) = 0.319								
a. Dependent Variable: PE								

Multiple regression analysis was performed to determine the effect of PEOU and PU and PE on eSAT and the result is presented in Table 4. Based on Table 4, the positive effects of PEOU (β 0.084, t 2.499, Sig. 0.013), PU (β 0.320, t 7.368, Sig. 0.000) and PE (β 0.530, t 12.560, Sig. 0.000) on eSAT are observed. Therefore, H3a, H3b and H3c are accepted. In addition, the R_2 value also shows that 73.4% of the variation of eSAT is explained by the variation in PEOU, PU and PE.

Table 4: Results of the effects of PEOU and PU and PE on e-SAT

Coefficients[a]

Model		Unstandardized Coefficients		Standardized Coefficients	t	Sig.	Collinearity Statistics	
		B	Std. Error	Beta			Tolerance	VIF
1	(Constant)	.227	.115		1.965	.050		
	PEOU	.096	.039	.084	2.499	.013	.614	1.628
	PU	.345	.047	.320	7.368	.000	.370	2.704
	PE	.481	.038	.530	12.560	.000	.393	2.542
F Statistic = 349.665								
R Square (R^2) = 0.734								
a. Dependent Variable: eSAT								

5.0 Discussions

This study confirmed the appropriateness of applying utilitarian aspects (PEOU and PU) in predicting users' e-satisfaction (eSAT) in the usage of Facebook (a hedonic information system) among Malaysian young users. In this study, PEOU was found to have positive impact on PU (β 0.599, t 14.637, Sig. 0.000) in the usage of Facebook. The positive effect of PEOU on PU was consistent with the study of Sibona and Choi [7], Liao *et al.* [11], Kim *et al.* [25], Lane and Coleman [26]. It implies that the degree to which users' believe that using Facebook will be effortless is important in influencing their perceived performance of the site.

PEOU also exerted positive effect on PE (β 0.564, t 13.362, Sig. 0.000), denoting that ease of use in the usage of Facebook influenced the degree of fun and pleasure experienced by users. This result is consistent with what being addressed by Bruner II and Kumar [14] in the usage of internet devices, Gu *et al.* [28] in the context of multipurpose information systems and Moqbel [27] in the adoption of SNSs. It implies that the degree to which users' believe that using Facebook will be effortless is important in determining their personal pleasure in the usage this SNS.

In addition, this study revealed that PEOU and PE of TAM and PE significantly and positively affect eSAT of the users. This shows that both utilitarian and hedonic aspects are important to satisfy users' needs and wants in the context of Facebook. These results are aligned with the work of Sibona and Choi [7] who found PEOU and PU increased eSAT in the usage of Facebook in United States and the study of Lin and Lu [6] who highlighted the positive effect of PE on eSAT in the usage of various SNSs in Taiwan. The positive impacts of PEOU and PU TAM and PE on eSAT implied that when:

- Measuring e-satisfaction of SNSs, it is important to include both utilitarian and hedonic aspects of information systems. The utilitarian aspect is expected to facilitate efficiency and productivity in using SNSs, and this will ensure users' experiencing pleasure and enjoyment.
- Developing or enhancing a SNS, it is important to include both utilitarian aspect (PEOU and PU) and hedonic aspect (PE) in order to increase users' eSAT. For instance, the service providers or developers of SNSs should ensure their sites are easy to use, navigate and communicate with, in which they will contribute to the productivity in the usage of SNSs. However, the primary focus of the service providers should still remain on hedonic aspect since PE was found to have stronger effect than PEOU and PU. Therefore, the developers of SNSs should always upgrade their sites to provide interesting and up-to-date features such as applications and games that could enhance users' experience and enjoyment.

6.0 Conclusion

The findings of this research revealed that both utilitarian and hedonic aspects of information systems, such as SNSs, are important in determining the e-satisfaction of users. PEOU and PU (TAM) of utilitarian information system and PE of hedonic utilitarian system are good predictors of e-SAT of SNSs. However, although all the results have been tested to be statistically significant, there are some limitations in this study. This research employed two variables (PEOU, PU) of utilitarian information system in determining the users' e-satisfaction in the usage of Facebook. These PEOU and PU are related to how users perceived the productivity of SNSs. In the study of electronic service quality or website quality, productivity of a website is also influenced by the functionality aspect of the site, for example efficiency, fulfilment and system availability. For instance, efficiency affect website speed and website proper work and no interruption in the usage of SNSs of system availability ensure continuity of users' business, and therefore, may influence their perceived productivity. These factors may have significant effect on the utilitarian aspect of SNSs, and therefore, may influence users' e-SAT. Thus, the inclusion of these factors in the utilitarian aspect of SNSs would be interesting to research.

Website aesthetic value may be considered to construct PE as it may influence users' feeling of entertainment, and therefore, it could mediate the relationship between PE and e-SAT in the usage of SNSs. The aesthetic value of a website deals with the website design. How attractive is a website design, for instance the appearance and beauty of the site and its visually appealing in term of graphics, layout used in the web to the users may attract them to use a SNS. Thus, the inclusion of this factor in the hedonic aspects of SNSs would be interesting to research.

References

[1] Boyd, D. M., and Ellison, N. B. (2007). Social network sites: Definition, History and Scholarship. *Journal of Computer-Mediated Communication*, 13(1), article 11.

[2] Zhong, B., Hardin, M., & Sun, T. (2011). Less effortful thinking leads to more social networking? The associations between the use of social network sites and personality traits. *Computers in Human Behaviour*, *27*(3), 1265-1271.

[3] Facebook Statistics (2013). About Facebook Statistics. Retrieved May 5, 2013, from https://www.facebook.com/pages/Facebookstatistics/119768528069029?fref=ts#

[4] Socialbakers (2013). Malaysia Facebook Statistics. Retrieved May 5, 2013, from http://www.socialbakers.com/facebook-statistics/malaysia.

[5] Shiau, W. L., & Luo, M. M. (2010). Continuance Intention of Blog Users: The Impact of Perceived Enjoyment and User Involvement.

[6] Lin, K.-Y., & Lu, H.-P. (2011). Why people use social networking sites: an empirical study integrating network externalities and motivation theory. Computers in Human Behavior, 27(3), 1152–1161.
[7] Sibona, C., & Choi, J. H. (2012). Factors Affecting End-User Satisfaction on Facebook. In *Sixth International AAAI Conference on Weblogs and Social Media*.
[8] Taghizadeh, H., Taghipourian, M. J., & Khazaei, Amir. (2013) The effect of customer satisfaction on word of mouth communication. Research Journal of Applied Sciences, Engineering and Technology 5(8): 2569-2575.
[9] Ong, C. S., & Day, M. Y. (2010). An integrated evaluation model of user satisfaction with social media services. In *Information Reuse and Integration (IRI), 2010 IEEE International Conference on* (pp. 195-200). IEEE.
[10] Devaraj, S., Easley, R., & Crant, J. (2008). How does personality matter? Relating the five-factor model to technology acceptance and use. *Information Systems Research*, 19 (1), 93-105.
[11] Liao, C. H., Tsou, C. W., & Shu, Y. C. (2008). The Roles of Perceived Enjoyment and Price Perception. *International Journal of Business and Information, 3*(1).
[12] Leng, G. S., Lada, S., Muhammad, M. Z., Ibrahim, A. A. H. A., & Amboala, T. (2011). An exploration of social networking sites (SNS) adoption in Malaysia using Technology Acceptance Model (TAM), Theory of Planned Behavior (TPB) and intrinsic motivation. *Journal of Internet Banking and Commerce, 16*(2), 1-27.
[13] Van der Heijden, H. (2004). User acceptance of hedonic information systems. *MIS quarterly*, 695-704.
[14] Bruner II, G. C., & Kumar, A. (2005). Explaining consumer acceptance of handheld Internet devices. *Journal of Business Research, 58*(5), 553-558.
[15] Curran, J. M., & Meuter, M. L. (2007). Encouraging existing customers to switch to self-service technologies: put a little fun in their lives. *The Journal of Marketing Theory and Practice, 15*(4), 283-298.
[16] Wolfinbarger, M., & Gilly, M. C. (2011). Shopping online for freedom, control & fun. *California Management Review*, 43(2), 39-51
[17] Kwon and Wen, (2010). An empirical study of the factors affecting social network service use. Journal Computers in Human Behaviour, Volume 26 Issue 2, March, 2010 Pages 254-263.
18] King, W. R., & He, J. (2006). A meta-analysis of the technology acceptance model. *Information & Management, 43*(6), 740-755.
[19] Ernst, C. P. H., Pfeiffer, J., & Rothlauf, F. (2013). Hedonic and Utilitarian Motivations of Social Network Site Adoption. Working Paper, Johannes Gutenberg University Mainz.
[20] Pillai, A., & Mukherjee, J. (2011). User acceptance of hedonic versus utilitarian social networking web sites. Journal of Indian Business Research, 3(3), 180-191.Ramayah, T., & Ignatius, J. (2005). Impact of perceived usefulness, perceived ease of use and perceived enjoyment on intention to shop online. *ICFAI Journal of Management, 3*(3), 36-51.
[21] Lu, H. P., & Hsiao, K. L. (2010). The influence of extro/introversion on the intention to pay for social networking sites. *Information & Management, 47*(3), 150-157.
[22] Davis, F.D. (1989). Perceived usefulness, perceived ease of use, and user acceptance of information technologies, MIS Quarterly 13(3), 319-340.
[23] Koch, S., Toker, A., & Brulez, P. (2011). Extending the technology acceptance model with perceived community characteristics. *Information Research, 16*(2), 16-2.
[24] Venkatesh, V., & Davis, F. D. (1996). A model of the antecedents of perceived ease of use: Development and test*. *Decision sciences, 27*(3), 451-481.
[25] Kim, Y., Kim, M., & Kim, K. (2010). Factors influencing the adoption of social media in the perspective of information needs. Retrieved from University of Illinois, Illinois Digital Environment for Access to Learning and Scholarship website: http://hdl.handle.net/2142/15025

[26] Lane, M., & Coleman, P. (2012). Technology Ease of Use through Social Networking Media. Journal of Technology Research, 1-12.
[27] Moqbel, M. (2012). *The Effect of the Use of Social Networking Sites in the Workplace on Job Performance*. PhD Dissertation. Texas A&M International University.
[28] Gu, J.-C., Fan, L., Suh, Y.-H., & Lee, S.-C. (2010). Comparing Utilitarian and Hedonic Usefulness to User Intention in Multipurpose Information Systems. [Article]. Cyber Psychology, Behavior & Social Networking, 13(3), 287-297.
[29] Oliver, R. L. (1981). "Measurement and evaluation of satisfaction processes in retail settings," *Journal of Retailing*, vol. 57, pp. 25-48.
[30] Su, Y. L., Lin, T. M., & Chang, S. T. (2013). Exploring the impact factor of positive word-of-mouth influences satisfied customers. Total Quality Management & Business Excellence, (ahead-of-print), 1-15.
[31] Nithinant Thammakoranonta, Chirasil Chayawan, and Nattapat Boonprakate (2011). A Study of Factors Affecting Social Network Used in Thailand. Retrieved May 7, 2013, from http://iceb.nccu.edu.tw/proceedings/APDSI/2011/web/session/astudyoffactorsaffectingsocialnetwork.pdf
[32] Gómez-Borja, M. Á., María-del-Carmen Alarcón-del-Amo and Carlota Lorenzo-Romero (2012). *African Journal of Business Management*, *6*(29), 8609-8619.
[33] Ribbink, D., van Riel, A. C., Liljander, V., & Streukens, S. (2004). Comfort your online customer: quality, trust and loyalty on the internet. *Managing Service Quality*, *14*(6), 446-456.
[34] Krejcie, R.V. and Morgan, D.W. (1970). Determining Sample Size for Research Activities, Education and Psychological Measurement. Vol. 30, pg. 607-610.
[35] Leng, G. S., Lada, S., Muhammad, M. Z., Ibrahim, A. A. H. A., & Amboala, T. (2011). An exploration of social networking sites (SNS) adoption in Malaysia using Technology Acceptance Model (TAM), Theory of Planned Behavior (TPB) and intrinsic motivation. *Journal of Internet Banking and Commerce*, *16*(2), 1-27.
[36] Hise, R and Szymanski, D. (2000). E-satisfaction: An initial examination. *Journal of Retailing*, **76**(3), 309-322.
[37] Erin Cho 1 and Youn-Kyung Kim, "the Effects of Website Designs, Self-Congruity, and Flow on Behavioral Intention", International journal of Design. Vol. 6, No. 2, 2012.
[38] Song, Ji Hee and George M. Zinkhan, "Designing Successful Web Sites: Some Insights from IS and Marketing Theory", Enhancing Knowledge Development in Marketing, 14 (R.B. Money and R. L. Rose, Eds.). Chicago: American Marketing Association, 236-242, 2003.
[39] Zakuan N.M., Yusof, S.M. and T. Laosirihongthong, (2008) "Reflective review of relationship between TQM and Organizational Performance", *Proceeding of the 4th International Conference on Management of Innovation and Technology, ICMIT*, pp. 444-448

MSWT-01, an alternative in combining Production Based Education (PBE) and student CSR program in Polman

Gamawan Ananto[1], Albertus B Setiawan[2] and Darman MZ[3]
Manufacture Engineering Dept, Bandung State Polytechnic for Manufacturing, Jl. Kanayakan 21 Bandung 40135, INDONESIA

Email: gamawan@polman-bandung.ac.id

Abstract. MSWT-01, Mobile Surface Water Treatment, producing $1m^3$ per hour, is an alternative for providing clean water in flood disaster areas, and was developed at Bandung State Polytechnic for Manufacturing (Polman) as a part of institution research project. The combination of cartridge or membrane technology such as carbon block, MF, UF and filtration media is used for this machine, instead of coagulation-flocculation with chemical addition, due to emergency purposes related with its treatment processing time. The idea is that MSWT could be combined with Production Based Education (PBE) concept in Polman as a vocational education institution and students 'CSR', students social activities. With the number of implementation trials in real flood area condition, MSWT will be developed further based on the technical output result. The manufacturing process for improving or adding necessary features could be implemented as a student's project in PBE system. This might be an ideal combination alternative for such vocational institution that students get the product media for their PBE program and implement their work as a defined social activity. They will learn and experience related technical matters and more social interactions with the people and other disaster stakeholder as well.

1. Introduction

Since there are a lot of places in Indonesia with potential disaster that flood might be happened frequently, clean water lacking is one of problems faced in such places due to its water sources contaminating. For providing clean water, various solutions always initiated by all related parties such as Government mitigation programs, CSR program from some companies, and survival actions of people around these unlucky places. MSWT-01, Mobile Surface Water Treatment, that made for $1m^3$ per hour capacity, is an alternative to provide clean water in flood disaster area, which developed in Bandung State Polytechnic for Manufacturing (Polman) as a part of institution research project. The combination of cartridge or membrane technology such as carbon block, MF, UF and filtration media is used for this machine, instead of coagulation-flocculation with chemical addition, due to emergency purposes related with its treatment processing time. As known, in general, the optimal holding time of flocculation and coagulation process for such capacity is about 30-40 minutes[1].

Polman, as a vocational education institution, implement practical program in its laboratory or workshop that had the aim to build students skill competencies. For those practical product media the institution might create a structured exercise that referred to students competency levels needed, execute/ fabricate the research project result from internal demand of institution, and could also be initiated from external inquiry through collaborating with industrial parties. Such collaboration concept between academia and industry called Production Based Education. Implementing real life industrial practice within education activities is considered as a catalyst in this collaboration[2]. This can help university or polytechnic to achieve student competencies by providing exact learning media for always relevant to the industrial need purpose. The application of technology for industrial service activities in the laboratory/ workshop will help to validate an education process, since products/ services are always tried to meet the industrial need[3].

From other point of view, students should learn things related to organization, team work and social interactions. In industries, there is a concept called Corporate Social Responsibility (CSR). This concept has evolved since its establishment in the 1950's. Bowen (1953) was the first to propose a definition for the construct[4]. Social responsibility in the higher education sector is not so different from the corporate social responsibility. Both universities and corporations have to provide sound responses to new realities of environment and society. The primary role of universities is to serve the society within it functions[5].

The idea is that MSWT could be combined with PBE concept and students 'CSR' in Polman. With the number of implementation trials in real flood area condition, MSWT will be developed further based on the technical output result. The manufacturing process for improving or adding necessary features could be implemented as a student's project in PBE system. This MSWT development and students 'CSR' combination through PBE system will enrich students insight in product variant. Therefore, this might be an ideal combination alternative for such vocational institution that students get the product media for their PBE program and implement their work for a defined social activity. They will learn more interactions with the people and other disaster stakeholder as well.

2. Methodology

The steps of MSWT-01 implementation as shown on Figure 1. It has to be prepared before taken for 'lab. scale' trial or simulation on the nearest river or conditioned water source. The preparation is including previous data that content technical matters, check the general condition of electrical and mechanical properness, its mobile wagon etc. If this simulation is OK it will be brought for operation on real location of disaster after necessary improvement(s), if any. In this stage, there will be more proper preparations needed such as spare components, supporting tools, electric power generator, vehicle for trailing the wagon out, organizing related parties on disaster area etc. Post operation, an evaluation should be initiated that is following with improvement planning, feature adding, standard maintenance procedure or other necessary actions.

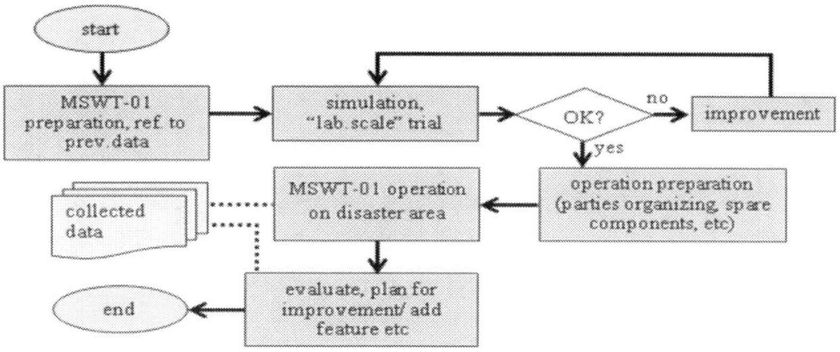

Figure 1. Implementation General Steps for MSWT-01

In vocational institution like Polman, necessary actions in improvement such as re-fabrication, features adding and other related tasks will be treated as 'research project' handling in PBE system, as mentioned on Figure 2. While job shop and batch products are running by students who assigned in each machine, products that resulted from research activity managed by small team that consist of Lecturer who had relevancy with the research theme and supported by students as the 'temporary' member of the team. The way of such team handling caused necessary improvement flow will be easier either it is still in design process/ engineering, fabrication, or looping back for re-design during in machining stage[6]. For MSWT case as a result of 'research project', its implementation could be social penetration, utilizing for benefit activity or other program.

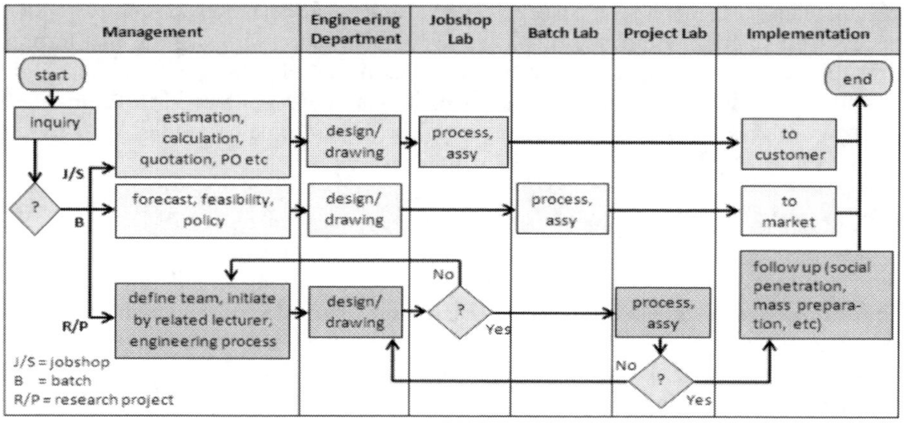

Figure 2. PBE General Order Flow for Job Shop, Batch and Research Project[6]

3. Discussion

3.1 Technical Aspects

MSWT-01 used the combination of cartridge or membrane technology such as carbon block, MF, UF and filtration media, instead of coagulation-flocculation, as stated. This is inspired by a number of references such as few water treatment technologies in Figure 3 [8] that shows how conventional filtration process may overcome undissolved compound content up to around 1μm dimension, whereas the most extreme condition such as seawater or brackish with salts content has to use RO (reverse osmosis) membrane. One situation may need conventional filtration only whereas the other could be added with MF for instance [9].

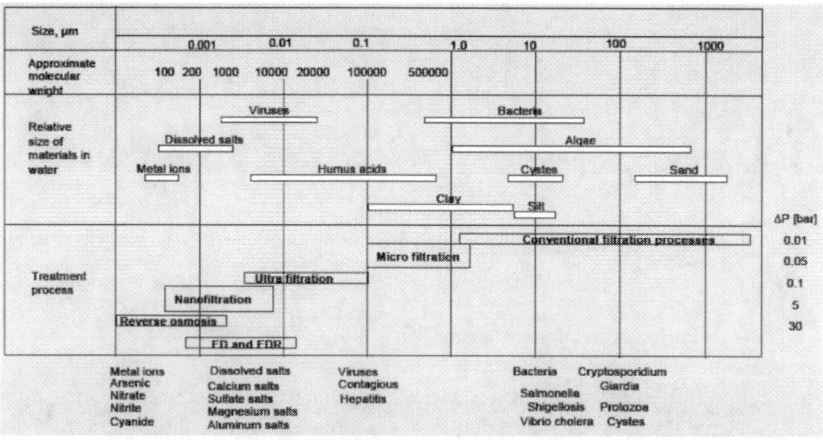

Figure 3. Water Treatment Process Technology Level[8]

On previous concept, there were number of trials that using oxidation and coagulation/ flocculation method. The result was absolutely not bad, but additional chemical solution and holding time are needed. Theoretically, the optimal holding time of flocculation/ coagulation process for such capacity is about 30-40 minutes[1] that to be considered due to emergency purpose for MSWT case. Therefore, standard cartridge is chosen for MSWT new concept instead of chemical way. The default features for treating the raw water source in MSWT-01 are carbon block (CB) cartridge, filtration media (active sand, active carbon & other mineral media) and micro filtration (MF) membrane. Referred to previous result of trials, this design is capable to produce deserved clean water that ready for cooking. In case the raw water condition on certain location is that worst or the asked demand is up to that high, the optional features such as ultra filtration (UF) membrane and ultra violet (UV) disinfection might be implemented or added, as indicated on Figure 4 [7].

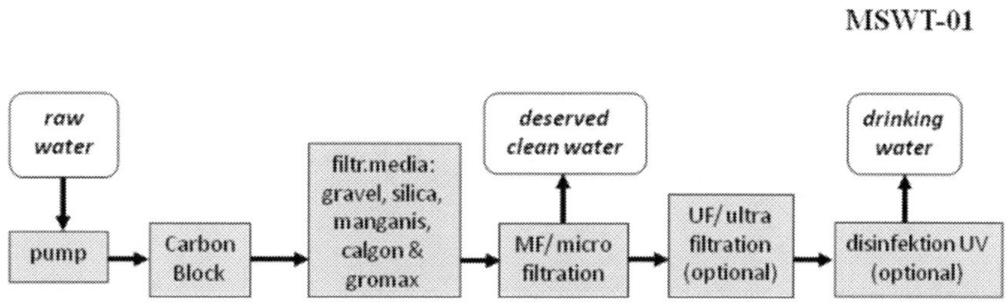

Figure 4. MSWT-01 Block Diagram [7]

Table 1 shows the result comparison between chemical (oxidation, flocculation/ coagulation) and CB (carbon block) & MF (micro filtration) cartridge method, which sand filtration are used for both. Since the Fe content in cartridge method result seemed not that proper, more trials still have to be initiated in order to show the stable and consistent result.

Table 1. Result Comparison, Chemical and Cartridge Method

Parameter	Unit	Std. max spec.	Source, raw water	Result, oxidation & floculation	Result, CB & MF cartridge
Color	TCU	15.0	35.90	6.00	3.00
Turbidity	NTU	5.0	199.80	2.00	0.80
Fe content	mg/L	0.3	8.96	0.20	0.56
Mn content	mg/L	0.4	0.46	0.30	0.22
Organic ($KMnO_4$)	mg/L	10.0	51.81	4.73	1.74
Hardness ($CaCO_3$)	mg/L	500.0	136.04	117.70	114.62

3.2 PBE (Production Based Education) and Students' Activities

Through PBE, Polman acts as an industry partner in components, tools and machine making. In general basic concept these activities are created to follow the present and updated technological needs of industries. The students practical program schedule is made structurally, combined with theory and other contents needed according the planned competency for each study program in 3 years (6 semesters) for Diploma-3, or 4 years (8 semesters) for Diploma-4.

As an example of Diploma-4 Study Program, Figure 5 shows how the education program is developed based in three phases arrangement: Problem-Based, Production-Based and Project-Based. Problem-Based is started using standard product as a structured job in order to train the student in basic competencies, fit to the standard with low-mid speed whereas ordered job in deepening a competences for more speed in delivering a product to the industry as a customer. Production-Based is operated in the real industry; students are working full time for 1 year according to daily industrial problem[3] that could be an external OJT (On the Job Training). Another case, one of Diploma-3 Study Program arranged these 2nd phase as an internal 'OJT' that students are doing the Industrial Practice in the campus.

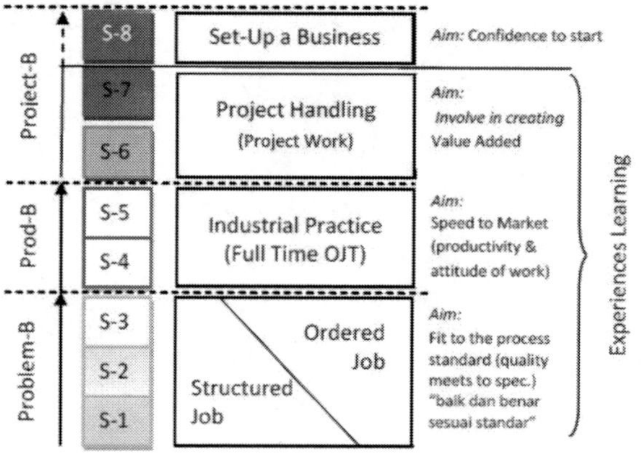

Figure 5. Education Management Program[3]

Called as '321' program idea, in this Diploma-3 students will get the first 3 semesters (sem) for basic matters needed including theory sessions, the next 2 semesters (sem 4 and 5) in integrated project program continually, then the last 1 semester for finalization and the additional enrichment items. Basically, instead of doing machining or other process respectively like sem 1-3, these sem 4 and 5 are defined to a few subprograms that won't be interrupted with any theory week nor surrounding from one kind of machine to another[6]. These subprograms will be accommodated in Engineering Department, Job Shop Lab, Batch Lab and Project Lab by turns.

	Month-1				Month-2				Month-3				Month-4				Month-5				Month-6			
Group A	BV	BV	JS1	JS2	JS2	D	D	D	M	P	P	P	P	P	P	P	P	A	A	PR	PR	PR	PR	E
Group B	JS2	JS2	BV	BV	PR	PR	PR	PR	D	D	D	M	P	P	P	P	P	P	P	P	A	A	JS1	E
Group C	D	D	D	M	P	P	P	P	P	P	P	P	A	A	PR	PR	PR	PR	JS1	JS2	JS2	BV	BV	E
Group D	PR	PR	PR	D	D	D	M	P	P	P	P	P	P	P	P	A	A	JS1	PR	BV	BV	JS2	JS2	E

Notes:

- BV = bench vice product, in Batch lab
- JS1 = Jobshop Lab, external customer's job
- JS2 = Jobshop Lab, external customer's job
- PR = Project Lab, internal customer's job
- D = design, drawing
- M = component & material preparation
- P = machining process
- A = assembly

Figure 6. 321" Practical Work Schedule For Semester 5 [6]

A simplified example program schedule in sem 5 is shown by Figure 6. One of subprogram is a 'project based' for each subgroup (A, B, C and D) that composed -in this case- for 6 students, since 1 group consists of 24 students. MSWT-01 could be done by one of such project based subgroup. When the machine is ready, related subgroup will be assigned to do the implementation on disaster location as far as their schedule program fitted with the disaster case, otherwise other student group should be assigned conditionally through extracurricular program that prepared as a student 'CSR' program, where they could experience all related technical matters and more social interactions with the people. A maintenance procedure could be initiated as well for this MSWT-01 periodically, through project based subprogram or part of it.

4. Conclusion

MSWT-01, Mobile Surface Water Treatment, is developed in Bandung State Polytechnic for Manufacturing (Polman). As an institution research project it has to be implemented as proposed, by utilizing on flood disaster with number of improvements and maintenance support. With the number of trials in real flood area condition, its further development based on the technical output result. The steps of implementation are; preparing all necessary things such as spare components, supporting tools, organizing disaster area related parties etc, and operation itself. Post operation, an improvement planning, feature adding, standard maintenance procedure or other necessary action has to be initiated based on obtained evaluation.

In Polman as a vocational institution that implement Production Based Education (PBE) system, necessary actions for MSWT-01 manufacturing, assembling & function tests could be treated as 'research project' handling. Such product that resulted from research activity will be managed by small team that consist of Lecturer with relevant research theme and supported by students as the 'temporary' team member. When the machine is ready, a student subgroup will be assigned to do the implementation on disaster location following their schedule program, or other student group should be conditionally assigned through extracurricular program that prepared as a student 'CSR' program. This might be an ideal combination alternative for such vocational institution for education that students get the product media for their PBE program and implement their work for a defined social activity. In addition, they will learn and experience all related technical matters and more social interactions with the people and other disaster stakeholder as well.

5. References

[1] Lindu M 2008 The Effects of Gradients Velocity and Detention Time to Coagulation-Floculation of Dyes and Organic Compound in Deep Well Water *Indo. J. Chem.* **8 (2)** p146-150

[2] Ilyas PI and Semiawan T 2012 Production Based Education (PBE): The Future Perspective of Education on Manufacturing Excellent *SciVerse ScienceDirect, Procedia-Social and Behavioral Sciences* **52** (2012) p5-14

[3] Nurdin M 2012 Technopreneurship Model Through Industrial Collaboration In Applied Technology Higher Education *IC-GWBT2012, Ahmad Dahlan University* p192-199

[4] Poulton E and Barnes L 2012 Corporate Social Responsibility: an Examination of Business Students' Perceptions of Materialism *Zenith, International Journal of Multidisciplinary Research* **Vol.2 Issue 6**, June, ISSN 2231 5780

[5] Burcea M and Marinescu P 2011 Students' Perception on CSR at The Academic Level case Study: The Faculty of Administration and Business, Universsity of Bucharest *AE (Amfiteatru Economic), Corporate Social Responsibility* **Vol. XIII no 29** p207-220

[6] Ananto G 2012 '321' Program In Polman For Better Operations Management Understanding *Advanced Material Research* **576** (2012) p685-689

[7] Ananto G, Setiawan AB and Darman MZ 2013 MSWT-01, Flood Disaster Water Treatment Solution From Common Ideas *IOP Conf. Series: Material Science and Engineering* **46** (2013) 012036 doi 10.1088/1757-899X/46/1/01//2036

[8] Heijman SGJ 2007 Nanofiltration and Reverse Osmosis **CT4471** *TU Delft*, p2.

[9] Boselovic L and Mc Clatchy 2009 Carbon wins contract for Phoenix water treatment *ProQuest document* **ID1923254051**, Washington: Calgon - Tribune Business News Dec 17

Causal Analysis of ISO/TS16949 Efforts in Automotive Industry

Norhayati Zakuan, Mohd Shoki Md Ariff , Mohd Norfian Alifiah
Faculty of Management,
Universiti Teknologi Malaysia, 81310 UTM Skudai, Johor, Malaysia

Shari Mohd Yusof, Muhamad Zameri Mat Saman
Faculty of Mechanical,
Universiti Teknologi Malaysia, 81310 UTM Skudai, Johor, Malaysia

E-mail: norhayatimz@utm.my

Abstract. The principal aim of this study is to determine the impact of ISO/TS16949 efforts on total quality management (TQM) and the ISO/TS16949 efforts relationship in Malaysian automotive industry. A survey method was employed to collect data using English version for Malaysian respondents. Exploratory factor analysis and reliability analysis empirically verified and validated the underlying items of TQM, ISO/TS16949 and organizational performance. Structural equation modeling was employed to test the proposed model. Data analysis revealed that TQM implementations have a strong and positive impact on organizational performance through direct relationship as well as through a mediating role of ISO/TS16949 efforts. From the research findings, it is viable to recommend that ISO/TS16949 should be incorporated with philosophy and methods of TQM to achieve better performance. This study proved to be valuable to academic researchers as well as automotive practitioners. It systematically analyzes the mediating effect of ISO/TS16949 efforts on TQM and organizational performance relationship.

Keywords – ISO/TS16949 efforts, Automotive Industry, Mediator, Organizational performance, Structural equation modeling and Total quality management.

1. Introduction

The implementation of quality management system has not occurred in the same pace in different regions of the world [1]. While early implementation started in Japan, it later shifted to the US, then Europe, and later to the developing countries. To compete in the global market, these countries, especially the late bloomers, need to implement quality management practices, tools, and techniques within all sectors of their industries [2]. Despite the huge number of publications and research on quality management, little empirical work has been carried out in developing countries, particularly in the ASEAN region [3,4] . Thus, this research contributes to the understanding of the differences that could exist in this part of the world with regard to quality management implementation.

Majority of existing research focuses on TQM implementation and performance relationships [5,6], whereas the role of ISO/TS16949 efforts on TQM and performance relations in the context of automotive industry is largely neglected. The practice of TQM also affects from the national level to the international level [6], which helps organizations to compete internationally and gain a competitive edge in the global market [7].

ISO/TS16949:2002 is an ISO technical specification that represents a comprehensive quality management system for the global automotive industry to achieve world class levels of product quality, productivity, competitiveness and continual improvements. It was clear from its inception that ISO/TS16949:2002 was different than its predecessor, QS 9000, for many reasons.

Previous studies have also failed to consider the significant potential of mediating variables such as ISO/TS16949 certification in automotive industry. Positive relationship (in the sense of certification leads to better performance) between TQM and organizational performance is often taken for granted even though this relationship has not been proven. Furthermore, critics who opined that ISO certification provide only "hollow achievement" [7] is gaining mileage. Whether ISO/TS16949 certification produces the expected benefits or it is just a hollow achievement, empirical evidence has yet to be found. At the very least it is important to assess the performance of organizations with ISO/TS16949 certification if indeed better than those of without certification.

The purpose of this paper is to propose a structural model that relates TQM ISO/TS16949 and organizational performance. It is an attempt to contribute to the extant literature by incorporating TQM practices and ISO/TS16949 efforts on organizational performance relationship within the context of an emerging automotive industry. By using structural equation modeling, the study analyzes both direct and indirect effects of each construct on organizational performance.

2. Methods

Mail survey was conducted to collect data in this study because of its relative simplicity in collecting large quantitative data. The survey kit consisted of a large (9" x 12") mailing envelope which include the survey instrument and cover letter stapled together, as well as a post-paid self-addressed envelope. The survey kit was mailed in various batches to the target respondents.

Second mail to non-responding respondents was done one month after the initial mails. To minimize costs, survey kits with reminder were sent to every alternate non-responding target on the mailing list. Post-paid post cards were sent to the remaining non-responding respondents as reminders requesting the recipient to complete and return the questionnaire. In the case where the original survey was lost and the recipients were willing to complete the survey, new survey kits were sent.

Even after the follow up mails were done, the response rate was still around 10% especially in Thailand. To increase the response rate telephone calls and follow-up fax transmission were made to persuade them to fill up the questionnaires. In Malaysia, the questionnaire was distributed to a total of 650 companies and 161 completed forms were received giving a response rate of 25%.

2.1. Treatment for missing data

Since the structural equation modeling (SEM) software was used to analyze part of the data, missing data became an issue. SEM requires that "complete data are required for the probability density and adjustment must be made to data sets that are incomplete" [8]. Missing data can exist in two forms, either due to omission during data entry process or accidental lack of response by the respondent. Different procedures were used to adjust to the missing data under each situation. The first procedure was to eliminate input errors on the part of the data entry by re-examining the original questionnaire and the data entry error. There were five cases in Malaysia dataset of this type.

The second procedure addressed the issue no response by respondents. In this study, if less than 5% of the data is missing, the missing data were filled in with the mean value [9], otherwise, the data were dropped from the study. For Malaysian respondents, there were 10 cases the missing data filled in with the mean value and four cases were dropped from the dataset. Validity and reliability
Exploratory factor analysis (EFA) with varimax rotation was performed on the TQM constructs, organizational performance (OP) measures and ISO/TS16949 efforts. At a minimum, a 0.4 loading of each item on its respective factor were considered adequate for that factor and a minimum of 0.5 loading for items that cross-load on multiple factors were considered adequate [10]. The EFA of 53

items of TQM constructs yielded eight factors explaining 57.92% of the total variance. The result indicated that eight TQM constructs have been identified with 40 items as compared to the original questionnaire which had 53 items. Subsequently, the EFA of 14 items of organizational performance were loaded on two factors explaining 51.70% of the total variance. No items were recommended to be omitted. Finally, the EFA of 34 items of ISO/TS16949 were loaded on four factors explaining 61.58% of the total variance. No items were recommended to be omitted from the original questionnaire.

The Cronbach's α measure of reliability for TQM constructs and organizational performance was between 0.851 and 0.730. According to Aamadi [11], the generally accepted minimum value of α is 0.70, however, Nunnally [12] allowed a slightly lower minimum limit, such as 0.6 for exploratory work involving the use of newly developed scales. Since the Cronbach's α value for each factors was above 0.70, all factors were accepted as being reliable for the research. Table 2 shows the results of EFA and reliability analysis.

Table 2. EFA and reliability analysis of the TQM, OP and ISO/TS16949

Factors	Number of items	First Eigen-value	Percentage of variance explained	Cronbach α
TQM		21.03	57.92	
QL	7			0.851
CFS	5			0.795
QIA	4			0.802
HRD	4			0.824
SPM	5			0.799
SQM	5			0.836
QR	5			0.746
QA	5			0.730
OP		6.09	51.70	
LS	6			0.816
BR	8			0.842
ISO/TS		15.54	61.58	
CP	3			0.769
MTT	19			0.946
ImpAct	8			0.912
IntAud	4			0.858

3. Results and Discussion

3.1 Mediating effect of ISO/TS16949 efforts in Malaysia

Figure 1 shows both inner and outer regression weights for the structural relationships between TQM, ISO/TS16949 efforts and organizational performance, based on structural equation modeling for Malaysia automotive industry (n = 157).

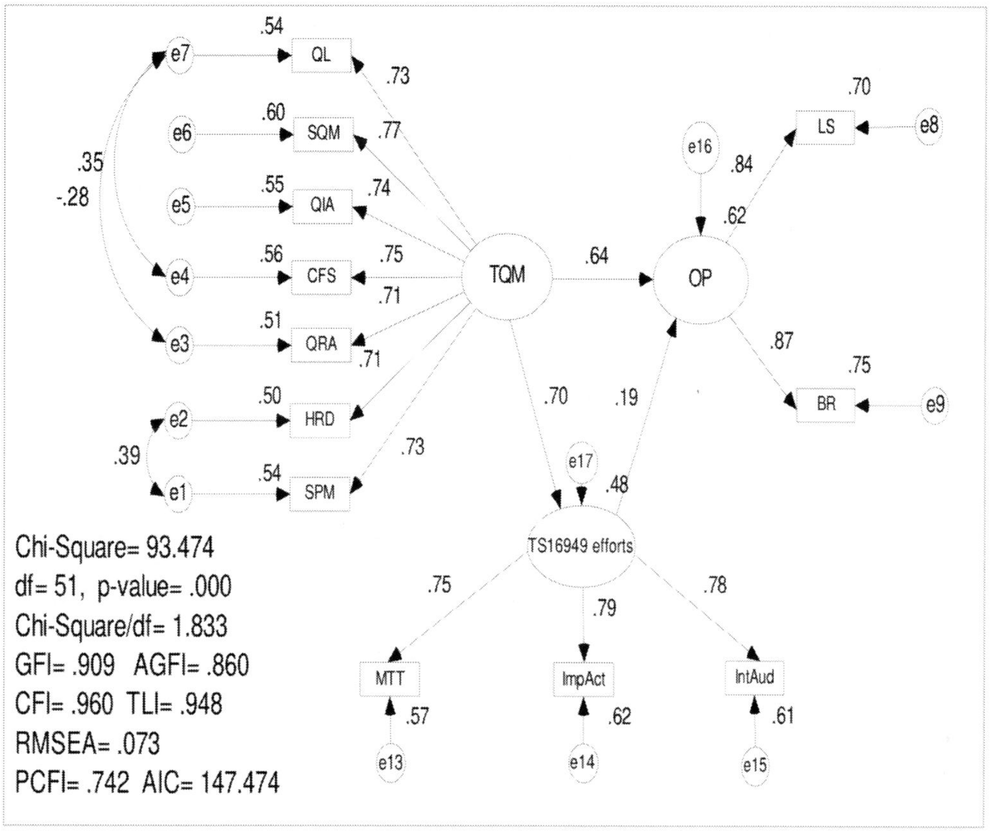

Figure 1. Inner and outer regression weights for the structural relationship between TQM and OP with a mediation of ISO/TS16949 efforts for Malaysia automotive industry

The goodness-of-fit indices for the structural model (χ^2/df = 1.833, GFI = 0.909, AGFI = 0.860, CFI = 0.960, TLI = 948, and RMSEA = 0.073) were well within the generally accepted limits, indicating a good fit to the data. The results showed that, the standardized regression weight for *H1a* was found to be 0.644 and significant ($p < 0.001$). This result provided support for the H1a which states that TQM implementation has a direct and strong impact on organizational performance of Malaysian automotive industry.

The standardized regression weight for the hypothesized relationship between TQM implementation and ISO/TS16949 was found to be positive (0.696) and significant ($p < 0.001$), confirming *H2a* that TQM had a strong positive direct impact on the level of ISO/TS16949 efforts.

The standardized regression weight for the direct relationship between ISO/TS16949 efforts and OP was found to be positive (0.190) and insignificant with p-value of 0.085 which was more than 0.05. The result did not provide support hypothesis *H3a* that ISO/TS16949 efforts have a direct and strong impact on organizational performance.

Organizational performance is also indirectly affected by TQM through ISO/TS16949 efforts. In order to test whether ISO/TS16949 efforts are an important mediator of TQM with organizational performance (OP) relationship the rule of thumb was followed Hair [13], namely (i) IE < 0.085 => Non mediator, (ii) IE > 0.085 and IE ≈ DE => Partial Mediator (TQM -> OP relationship, $p < 0.05$) and (iii) IE > 0.085 and IE >>>DE => Total Mediator (TQM -> OP relationship, $p > 0.05$).

Table 3 shows the standard Indirect Effect (IE) of TQM to OP is 0.132 which was more than 0.085. Thus, ISO/TS16949 efforts mediate the relationship between TQM and organizational performance (OP). Since, the p-value for Direct Effect (DE) between TQM to OP was less than 0.05, the ISO/TS16949 efforts were considered as partial mediator. In conclusion, this finding provides support to *H4a* hypothesis: the impact of TQM implementation on organizational performance increases with a mediating role of ISO/TS16949 efforts in Malaysian automotive industry.

Table 3. Direct (DE) and Indirect Effects (IE) analysis for Malaysia automotive industry

	Std. Total Effect			Std. Direct Effect			Std. Indirect Effect		
	TQM	TS16949	OP	TQM	TS16949	OP	TQM	TS16949	OP
TS16949	0.692	0.000	0.000	0.692	0.000	0.000	0.000	0.000	0.000
OP	0.777	0.190	0.000	0.646	0.190	0.000	0.132	0.000	0.000

Note: Std. Total Effect = Std. Direct Effect (DE) + Std. Indirect Effect (IE)

The discussion thus far revealed that the structural models were valid for Malaysia. Table 4 shows that relationship between TQM and organizational performance for Malaysia automotive industries were found to be positive and significant, confirming that TQM implementation had a strong positive direct impact on the organizational performance. The direct effect relationship between TQM implementation and organizational performance found in this study corroborates the findings of some previous studies including Ahire and Golhar [14], [5] and [15]. Table 4 also presents the relationship between TQM implementation and ISO/TS16949 efforts. The findings indicate that TQM implementation had a positive and strong direct impact on ISO/TS16949 efforts.

Table 4. Inner regression weights between latent variables in the structural model for Malaysian automotive industries

Country	Hypothesis	Structural relationships	Standardized regression weight
Malaysia	*H1a*	TQM – OP	0.644*
	H2a	TQM – ISO/TS16949	0.696*
	H3a	ISO/TS16949 – OP	0.190
	H4a	TQM – OP (indirect relationship)	0.132*

Note: *$p < 0.001$

4. Conclusion

This study has provided an empirical justification for a research model that identifies the relationship among TQM implementation, ISO/TS16949 certification and organizational performance. Many studies have been performed to identify enabling factors for successful implementation TQM. However, very limited previous studies had investigated the relationships between TQM practices and organizational performance, especially amongst ASEAN countries. Data for the study were collected from a sample of 161 Malaysian automotive industries. Based on the exploratory factor analysis and reliability analysis, all factors were accepted as being valid and reliable for the research. The results of this study showed that ISO/TS16949 efforts partially mediate the relationship between TQM implementation and organizational performance in Malaysian automotive industry.

Acknowledgement

The authors would like to thank the Faculty of Management, Universiti Teknologi Malaysia, Johor Bahru, Malaysia.

References

[1] N.M. Zakuan, S.M. Yusof, T. Laosirihongthong (2008). Reflective review of relationship between total quality management and organizational performance, *Proceeding of the 4th IEEE International Conference on Management of Innovation & Technology (ICMIT2008).* pp. 444-448

[2] Iwaarden, J.V., Wiele, T.V.D. and Williams, R. (2006). A Management Control Perspective of Quality Management: An Example in The Automotive Sector. *International Journal of Quality & Reliability Management.* 23(1), pp. 102-112.

[3] N. Zakuan. M.Z. Mat Saman (2009). Lean manufacturing concept: The main factor in improving manufacturing performance – A case study. *International Journal of Manufacturing Technology and Management 17(4).* pp. 353-363

[4] Aarabi, M., Saman. M.Z.M., Wong, K.W., Azadnia, A.H. and Zakuan, N. (2012). A comparative study on critical success factors (CSFs) of ERP systems implementation among SMEs and large firms in developing countries. *International Journal of Advancements in Computing Technology 4(9).* pp. 226-239

[5] Hendricks, K.B. and Singhal, V.R. (2001). Firm characteristics, total quality management, and financial performance. *Journal of Operations Management.* 19, pp. 269-285.

[6] Prajogo, D.I. and Sohal, A.S. (2004). The multidimensionality of TQM practices in determining quality and innovation performance – an empirical examination. *Technovation.* 24, pp.443-453

[7] Jones, R., Arndt, G. and Kustin, R. (1997). ISO 9000 among Australian companies: impact of time and reasons for seeking certification on perceptions of benefits received. *International Journal of Quality and Reliability Management.* 14(6/7), pp. 650-660.

[8] Brown, R.L. (1994). Efficacy of the indirect approach for estimating structural equation models with missing data: A comparison of five methods. *Structural Equation Modelling.* 1(4), pp. 287-316.

[9] Tabachnick, B. and Fidell, L. (1996). *Using multivariate Statistics.* New York, HarpereCollins.

[10] Ngai, E.W.T., Cheng, T.C.E., and Ho, S.S.M. (2004). Critical success factors of web-based supply chain management systems: An exploratory study. *Production Planning and Control.* 15, pp. 622-630.

[11] Aamadi, A. (2002). *Mathematic and Statistic.* Tehran Universities Publications Centres: Tehran.

[12] Nunnally, J.C. (1978). *Psychometric Theory.* 2nd ed., McGraw-Hill, New York, NY.

[13] Hair, Jr., Anderson, J., R. Tatham and Black, W. (1998). *Multivariate data analysis with readings.* 4th ed. Prentice Hall, Englewood Cliffs: New Jersey.

[14] Ahire, S.L. and Golhar, D.Y. (1996). Quality management in large vs small firms, *Journal of Small Business Management.* 34(2), pp. 1-11.

[15] Demirbag, M., Koh, S.C.L., Tatoglu, E., and Zaim, S. (2006). TQM and market orientation's impact on SMEs' performance. *Industrial Management & Data Systems.* 106(8), pp. 1206-1228.

Consumer Perceived Risk, Attitude and Online Shopping Behaviour; Empirical Evidence from Malaysia

Mohd Shoki Md Ariff[1], Michele Sylvester[1], Norhayati Zakuan[1*], Khalid Ismail[2], Kamarudin Mat Ali[3]

1 Department of Business Administration, Faculty of Management, Universiti Teknologi Malaysia (UTM), Skudai, Johor, Malaysia
2 Universiti Pendidikan Sultan Idris, Malaysia
3 Lembaga Tabung Haji Malaysia

E-mail: norhayatimz@utm.my

Abstract. The development of e-commerce has increased the popularity of online shopping worldwide. In Malaysia, it was reported that online shopping market size was RM1.8 billion in 2013 and it is estimated to reach RM5 billion by 2015. However, online shopping was rated 11th out of 15 purposes of using internet in 2012. Consumers' perceived risks of online shopping becomes a hot topic to research as it will directly influence users' attitude towards online purchasing, and their attitude will have significant impact to the online purchasing behaviour. The conceptualization of consumers' perceived risk, attitude and online shopping behaviour of this study provides empirical evidence in the study of consumer online behaviour. Four types of risks - product risk, financial, convenience and non-delivery risks - were examined in term of their effect on consumers' online attitude. A web-based survey was employed, and a total of 300 online shoppers of a Malaysia largest online marketplace participated in this study. The findings indicated that product risk, financial and non-delivery risks are hazardous and negatively affect the attitude of online shoppers. Convenience risk was found to have positive effect on consumers' attitude, denoting that online buyers of this site trusted the online seller and they encountered less troublesome with the site. It also implies that consumers did not really concern on non-convenience aspect of online shopping, such as handling of returned products and examine the quality of products featured in the online seller website. The online buyers' attitude was significantly and positively affects their online purchasing behaviour. The findings provide useful model for measuring and managing consumers' perceived risk in internet-based transaction to increase their involvement in online shopping and to reduce their cognitive dissonance in the e-commerce setting.

Keywords: Perceived risks, Attitude, Online Shopping Behaviour

1.0 Introduction

The growth of Internet has increased the popularity of online shopping [1] and it is the third most popular Internet activity, following e-mail using/instant messaging and web browsing [2]. Online shopping allows people to buy anything at any time, thus making it the most flexible way of purchasing electronically. It was reported that Malaysia's online shopping market size was RM1.8 billion in 2013 and it is estimated to reach RM5 billion by 2015 [3]. According to Malaysian Communication and Multimedia Commission [4], online shopping was ranked in the 11th position (of 15 purposes) why Malaysian access internet. These data provide two implication in the study of e-commerce or specifically consumer online shopping behavior; (i) the growth of online market size is very promising in the Malaysia's e-commerce landscape; and (ii) the online purchasing is not a major factor driving Malaysian to access the internet, thus posing questions how to improve its ranking, as well as to research factors hindering them to do online shopping. The former provides huge market growth for online shopping while the latter drives many researchers to examine on how consumers' perceived risk should be managed to allow more buyers committing in online purchasing. Thus, consumers' online shopping behavior should be researched in relation to their attitude and perceived risks as this will lead to the more Malaysian doing online transaction and this will give significant impact to Malaysia e-commerce development.

In marketing literature, it was recognized that risk perceptions directly affect purchasing and purchase intention, i.e. when consumers' perceived risk is high, the possibility of consumer to buy or to repurchase online will be low. It shows that perceived risk will affect the attitude of online purchasing [5]. Negative effects from the perception of risk have also been found to have a negative impact on shoppers' attitudes towards online shopping [6, 7 and 8]. Previous researches indicated that attitude towards online shopping is a significant predictor of making online purchases [9]. Obviously, risk can be real and as long as it is real it will affect consumers' purchasing behavior [10]. Then, consumers' perceived risk of online shopping must be continuously examined to monitor the extent to which they exert negative impact to consumers' online attitude and shopping behavior as well as to avoid them experiencing post-purchased dissonance. Hence, consumers' perceived risk should be continuously researched so that they can be proactively managed and reduces, thus contributing to the increase in online shopping. Therefore, this research is carried out to answer to the two research questions; (i) Do perceived risks affect attitude of Malaysian customers in online shopping? and (ii) does consumers' attitude affects their online shopping behavior?

2.0 Literature Review

In general term, online shopping is a process of buying products or services through Internet. Online shopping involves a user accessing internet to search, select, buy, use, and dispose of goods and services, in satisfying his or her needs and wants. In accessing and making online buying, individuals encounter a host of risk. In fact, consumers perceive a higher level of risk when purchasing through Internet compared with traditional retail transactions [11]. According to Ko et al. [12], perceived risk is the possibility of loss in achieving a favorable outcome while consumers doing online shopping, indicating individuals feeling of uncertainty with the possibility of negative outcome in their online purchasing.

Since Bauer [13] introduced risk-taking behavior in marketing literature as a possible measure of consumer attitude towards a purchase, individuals' perceived risk has been defined in several ways, with considerable debate concerning the merits of each point of view [14]. For example, Cultural Theory or Cultural Theory of Risk [15] relates consumers' perceived risk to cultural adherence and social learning. According to Wildavsky and Dake [15] the cultural theory of risk is used to "*predict and explain what kind of people will perceive which potential hazards to be how dangerous*". Hence, the variables of perceived risk are linked to the attitude of the consumers. Thus, to study the types of risks based on the cultural of Malaysian, consumers' perceived risks towards online attitude is included to the research framework.

Theory of Planned Behavior (TPB) was developed by Ajzen [16] and it is originated from the Theory of Reasoned Action (TRA). Generally, TPB is used to forecast a person's intention towards online shopping. TPB relates the conceptually of attitude, subjective norms and perceived behavioral control toward the behavior intentions of consumers. In this theory, attitude is determined by a behavioral belief about performing a particular behavior and it is one of the three factors in TPB's model. TPB theory stated that customer's behavioral attitude will directly influence the intention of online shopping. Past researchers, such as Pavlou and Fygenson [17], examined consumer's adoption of e-commerce by applying TPB and they found that the TPB was able to predict attitude and behavioral intention of consumers. Thus, attitude and online shopping behavior are added to the framework of this research.

Generally, there are many types of perceived risk experiencing by consumers in online shopping. Online shoppers reported different types of fear when they do online shopping. The levels of fears are listed in Table 1 [18].

Table 1: Level of fear according to different type of risks when consumers doing online shopping

Rank	**Type of risk**	**Observations**
1	Financial risk	The highest level of fear of respondents is the chances of suffering a financial loss due to fraud of credit card.
2	Performance risk	The second level of fear is that the product performance is not as expected.
3	Time risk	Similar to performance risk, fear of consumers about the product doesn't perform as described and caused the longer time of returning and waiting processes for the product to be delivered.
4	Delivery risk	Fear of consumers about the delivery will not be performed on time or that the product is damaged during the process of delivery.
5	Privacy risk	Fear of consumers about the privacy of their personal data and credit card details which may be misused by the seller.
6	Psychological risk	Fear of consumers regarding the doubts of e-Transactions, especially when the product is expensive or urgently needed.
7	Social risk	The lowest level of fear goes to social risk. It is about the fear of fraud transactions which may cause their referent groups (family members of friends) to blame them for making a bad decision.

Review of past researches on consumers' perceived risk of online shopping (2003–2012) is presented in Table 2. This review showed that, among the six types of perceived risk, four risks - financial risk, product risk, convenience risk and non-delivery risk - were frequently used in the study of risks in electronic mediated environment. Therefore, these risks are proposed as independent variables in the conceptual framework of this study.

Table 2: Summary of Previous Researches (2003-2012) on Dimensions of Perceived Risks in Online Shopping

Dimensions of perceived risk Researchers	Product/ performance Ri	Social Risk	Convenience Risk/Time risk	Non-Delivery Risk	Privacy Risk	Financial Ri
Swinyard and Smith (2003)	√		√			√
Hanjun *et al.* (2004)	√	√	√			√
Naiyi (2004)	√			√	√	√
Forsythe *et al.* (2006)	√		√	√		√
Javadi *et al.* (2012)	√		√	√		√
Zhang *et al.* (2012)		√	√	√		

Product risk

Online purchasing is a non-store shopping, so it is difficult for consumers to examine quality of physical goods because they can only rely on limited information, sketches, visuals, graphics, and pictures shown on the computer screen. As a consequence, the purchased products may not function as they are claimed to be. Product risk is generally about a purchased product may fail to function as originally expected [19]. It is related to the product performance risk, which is a loss for customers when a brand or product does not perform as expected due to the shoppers' inability to accurately evaluate the quality of the product online [20]. Several studies have shown that product or performance risk, or the fear of product not functioning or performing as expected is increasing in online environment [21]. In short, the incapability of touching, feeling, testing or trying the products before they are purchased are the primary concerns when purchasing online, and those concerns will increase the product or performance risk perceived [22].

Convenience risk

Convenience risk is about consumer perceptions of risks that the purchased product will take a lot of time and effort to repair and adjust before it can be used [23]. When consumers'' perceived convenience risk is high, they feel that it is very troublesome for them to perform certain online purchasing behaviour.

Financial risk

Consumers may feel worry about the online security with respect to using their credit cards and disclosing personal information. So, even though customers order goods online, but majority of them prefer alternative method of payments, such as cash at delivery, bank account transfers and Pay Pal other than using their credit cards. In other situation, consumers also fear that certain e-commerce websites are not secured enough and need constant reassurance. Previous Studies indicated that fear of credit card fraud represents one of the most invoked concerns when buying online [22].

Non-delivery risk

Non-delivery risk is one of the biggest worried when customers decided to buy products online. Dan *et al.*, [24] defined non-delivery risk as the potential fail of delivery which caused by goods lost, goods damaged and they were sent to the wrong place after confirmation of online order. According to Naiyi [25], consumers are concerns with the delivery process, for examples the product may get damaged during transportation, delivered to a wrong address, or in some cases, delayed. According to Claudia [18], consumers fear about delivery will be delayed due to various circumstances, such as the delivery company won't deliver the purchased products within the time frame agreed with customers. The feeling of fear that the products may be damaged during handling and transporting them to the consumers is also part of non-delivery risk.

Consumer Attitude and Online Shopping Behaviour

Attitude in term of online shopping is defined as consumer's positive or negative feelings when they are about to make the purchasing decision [26]. Attitude of customer is frequently related to emotion and according to the model of attitude change and behavior (e.g., Fishbein and Ajzen, 1975, [27], consumer attitudes are affected by their purchasing intention. The relationship between intention and behavior is based on the assumption that consumers attempt to make rational decisions based on information available to them. Therefore, an individual's behavioral intention to perform or not to perform certain behavior is the immediate determinant of that person's actual behavior [28].

Mitchell [29] suggested that perceived risk is a measure of expected dissatisfaction with a buying decision, based on the purchase goals of the buyers, and therefore, it is a powerful index for explaining consumer behavior since consumers are more often motivated to minimize potential failure than to pursue purchasing success. The reason is because perceived risk will directly influence customer's attitude and perceptions. Barnes *et al.*, [30] stated that perceived risk reduces the willingness of consumers to buy goods over the internet. Specifically, previous research indicated that attitude and purchase intention is negatively influenced by the perceived risk associated with the purchase [31]. If perceived risk is high then customers will have negative attitude [30].

Attitude is predicted to have a direct significant impact on online shopping behavior based on the studies of Chai and Pavlou [32], and George [33]. Thus, for those who have a positive attitude about online shopping will drive them to purchase online. Previous researches indicated that attitude towards online shopping is significant predictor of making online purchases [9] and purchasing behavior [9 and 33].

Online shopping behavior is a type of attitude which involve consumers browsing websites to search, select and purchase of goods and services, in fulfilling their needs and wants. It involves the reaction and choice of consumer decision making. Pavlou and Fygenson [17] stated that consumer intentions to use Internet as a shopping channel is a significant predictor of consumers' actual participation in online transactions. Generally, positive online shopping behaviour will lead to the successfulness of e-commerce transaction, such as online shopping.

3.0 Methodology

Review on the development of online shopping in Malaysia indicated that there is a need to study on how consumers' perceived risk and its effect on their attitude as well as the impact of consumers' attitude on their online shopping behavior. Generally, the conceptual model of this study was based on the integration of cultural theory or cultural theory of risk [15] and TPB [16]. In this study, perceived risk is independent variable, attitude (dependent variable/independent variable) and online shopping behavior (dependent variable). A questionnaire was used to measure the following variables of the study:

- Financial risk (three items), product risk (three items), and convenience risk (four items) were developed based on the work of Swinyard & Smith [34], Forsythe *et al.* [35] and Javadi et.al. [36]. Three items of non-delivery risk was adapted from Forsythe *et al.* [35].
- Consumer attitude with three items used was adopted from George [33].
- Four items of Online Shopping Behavior used were adopted from the work of Karayanni [37] and Forsythe *et al.* [35].

All the questions were measured using five point Likert scales of 1= Strongly disagree, 2= Disagree, 3= Neither Agree Nor Disagree, 4=Agree, and 5= Strongly agree.

The population of this research is the online shopping consumers of a Malaysia's largest online marketplace for online shoppers. This online seller is ranked 1st in Top 30 Local Websites accessed by Malaysians. In addition, for the Top 30 Combined Websites (International and Local) accessed by Malaysians, this site was ranked 3rd after Facebook.com and Google.com.my. The result of this ranking was collected from 25.57 million unique browsers on websites in Malaysia as reported by Malaysian Digital Association (MDA) in February 2012.

In multivariate research, the sample size required should be 5 to 10 time of variables for 10% and 5% margin error [38]. In this study, the total number of questions is 20, thus a minimum of 200 questionnaires are needed for 5% margin error and 100 questionnaires are required for 10% margin error. Using convenience sampling procedure, 300 completed questionnaires were collected. The reason of using convenience sampling is because it is a method of selecting volunteering respondents by depending on their availability and willingness to fill-up the questionnaire. At the same time, it is easy and proper for the researchers to access the respondents through Facebook page of the online seller site. The researcher had used web-based survey by creating the questionnaires using Google Form at https://docs.google.com/forms/d/122iVGBP9pA-ThhlVFveIFHZ0uDF3rvVZixcysv-ZVw/formResponse. The link of questionnaires was shared on the social media (Facebook) page of this site in order to find the right target of respondents. The reason of choosing social media is because currently the fan page of this site in Facebook is liked by 1,129,219 people. In addition, a filter question was used to confirm whether the respondents have the experience of online shopping at this site. In order to make sure the respondents are the online shopping consumers of this site, only respondents who are ever experience shopping at this site can proceed with the whole questionnaire. The data collected was kept automatically in Google spreadsheet and finally exported to Statistical Package for Social Science (SPSS) for the process of analysis.

4.0 Result and Analysis

For sampling adequacy, the values of Kaiser-Meyer-Olkin (KMO) for Perceived Risks, Attitude and Online Shopping Behavior were 0.726, 0.814 and 0.690 respectively (Table 3 and 4). The KMO measures of the sampling adequacy test and Bartlett's test of Sphericity were performed to confirm the suitability of the data for factor analysis. A high KMO value (close unity or > 0.6) and small p-value for the Bartlett's test (p-value < 0.05) were desired [38]. Since all KMO values were above 0.5 supported by Bartlett's test of Sphericity of 0.00, the factor analysis was then performed. Factor analysis is a process by which large clusters and grouping of data are replaced and represented by factors in the equation. It was applied as a data reduction or structure detection method, which can be used to identify the hidden dimensions or constructs which may or may not apparent from direct analysis. The results of Total Variance Explained of Confirmatory Factor Analysis indicated that for perceived Risk, the cumulative percentage was 71.499% (Table 3). Four factors, namely product risk, financial, convenience and non-delivery risks were identified and all items in these risks were retained and accepted based on the results of component matrix with factor loading of more than 0.5.

Table 3: Rotated Factor Matrix of Customers' Perceived Risk (PR – Product Risk, CR – Convenience Risk, FR – Financial Risk and NR – Non-delivery risk) of Online Purchasing

Code	Variables		Factors			
PR1	I will get what I ordered through this site		-.087	.015	**.871**	.071
PR2	I will not receive malfunctioning merchandise from this site		-.055	-.010	**.890**	.041
PR3	It is easy to judge the quality of merchandise over this site		.226	.013	**.836**	.000
CR1	I can examine the product if I shop at this site		**.785**	-.004	.167	-.046
CR2	If I shop at this site, I can wait till the product arrives		**.860**	-.040	-.024	-.031
CR3	It is easy to cancel orders with this site		**.858**	-.096	-.023	-.038
CR4	I won't have problem in returning product if I shop at this site		**.646**	.038	-.066	-.338
FR1	I feel that my credit-card details won't be compromised and misused if I shop at this site		-.076	.186	.145	**.770**
FR2	I will not get overcharged if I shop online as this site has my credit-card info.		-.016	.022	.063	**.871**
FR3	I feel my personal information given for transaction to this site will not compromise to third party.		-.194	.103	-.092	**.704**
NR1	I will receive the product ordered through this site		.005	**.849**	-.025	.167
NR2	I prefer this site because of availability of reliable & well-equipped shipper.		-.030	**.899**	-.015	.050
NR3	Products purchased at this site will be timely delivered		-.071	**.870**	.057	.081
	Total variation explained by these factors		71.499%			
	Kaiser-Meyer-Olkin Measure of Sampling Adequacy		0.726			
	Barlett's Test of Sphericity:					
	Approc. Chi-Square	D.F.	Significance			
	1551.240	78	0.000			

For attitude, the cumulative percentage of the total variance explained was 69.470%, and all items were retained in one dimension, i.e. consumer attitude, with factor loading of >0.5 (Table 4). For online shopping behavior, only one dimension has been extracted with total variance explained of 53.890 (Table 4). All of its items were accepted based on the results of component matrix with factor loading >0.5.

Further, the researchers performed reliability test for each dimension of perceived risk, attitude and online shopping behaviour. Cronbach's Alpha was often used for reliability coefficient assessing the consistency of the entire scale. Hair *et al.* [38,41] suggested that reliability estimated between 0.6 and 0.7 represent the lower limit of acceptability for reliability estimation. Cronbach's alpha value that is greater than 0.6 is considered as sufficient. In this study, the results of Cronbach's alpha values for all the dimensions are more than 0.7 (Product risk – 0.839; financial risk – 0.720; convenience risk –

0.812; non-delivery risk –0.854; attitude – 0.777; and online shopping behaviour – 0.722). Thus, construct of all the components used in this study are deemed reliable and valid for further analysis.

Table 4: Rotated Factor Matrix of Attitude (A) and Online Shopping Behavior (OSB)

Code	Variables	Factor	Code	Variables	Factor
A1	Shopping at this site is easy	**.781**	OSB1	I shop at this site as I can shop in privacy like at home	**0.732**
A2	Shopping at this site is fun and I enjoy it	**.858**	OSB2	I shop at this site because it provides me a broader selection of products.	**0.853**
A3	I prefer to buy at this site because it provides me with quality of information	**.688**	OSB3	This site gives me better control on my expenses.	**0.616**
			OSB4	Shopping at this site gives facility of easy price comparison.	**0.716**
	Total variation explained by this factor	**69.470%**		**53.890%**	
	Kaiser-Meyer-Olkin Measure of Sampling Adequacy	0.814		0.690	
	Barlett's Test of Sphericity:				
	Approc. Chi-Square / D.F. / Significance			Approc. Chi-Square / D.F. / Significance	
	1679.654 / 78 / 0.000			243.268 / 6 / 0.000	

To examine the effect of consumers' perceived risks on their attitude, multiple regressions were performed, and the result is presented in Table 5. Multiple regressions is a statistical technique to form a mathematical regression which primarily wanted to relate a dependent variable (Attitude) to more than one independent variables (consumers' perceived risk of online shopping). The result indicated that product risk (β -.161, t -3.005, Sig. 0.002), financial risk (β -.141, t -2.524, Sig. 0.012) and non-delivery risk (β -.110, t -2.035, Sig. 0.043) negatively affect attitude of online consumers. However, convenience risk was found to have positive effect on the attitude of consumers (β 0.304, t 5.577, Sig. 0.000).

Table 5: Results of the effects of Perceived Risk (Product risk (P), Convenience risk (C), Financial risk (F), Non-delivery risk (N)] on Attitude (A)

Coefficients							
Model	Unstandardized Coefficients		Standardized Coefficients	t	Sig.	Collinearity Statistics	
	B	Std. Error	Beta			Tolerance	VIF
(Constant)	3.624	.288		12.593	.000		
P	-.141	.046	-.161	-3.055	.002	.988	1.012
C	.292	.052	.304	5.577	.000	.930	1.076
F	-.142	.056	-.141	-2.524	.012	.880	1.137
N	-.093	.045	-.110	-2.035	.043	.944	1.059
Dependent Variable: Attitude (A)							
F Statistics = 16.888							
R Square(R^2) = 0.186							

It was found that attitude of consumers positively affect their online shopping behaviour (β 4.27, t - 8.161, Sig. 0.000). Result of linear regression, as shown in Table 6, indicated that consumers' online shopping behaviour was significantly and positively affected by their attitude towards online purchasing.

Table 6: Results on the effect of Attitude on Online shopping behaviour

Coefficients[a]							
Model	Unstandardized Coefficients		Standardized Coefficients	t	Sig.	Collinearity Statistics	
	B	Std. Error	Beta			Tolerance	VIF
(Constant)	2.138	.200		10.684	.000		

NEWSB	.446	.055	.427	8.161	.000	1.000	1.000
a. Dependent Variable: Online shopping behavior							
F Statistic = 66.608							
R Square (R^2) = 0.183							

5.0 Discussions

Theoretically, this research has confirmed three important findings. Firstly, product risk, financial risk, and non-delivery are negatively affected consumers' attitude of online shopping. This finding is consistent with several authors who found that consumer shopping behavior on the Internet, attitude toward usage behavior and intention to adopt e-commerce were affected by perceived risk [39 and 40]. According to Bhatnagar *et al.*, [20], the likelihood of purchasing on the Internet decreases with increases in product risk and this view is supported by this research. The negative effects of perceived risk have also been found to have a negative impact on shoppers' attitudes towards online shopping [7 and 8]. Financial risk with highest beta coefficient (β) -0.141 has the strongest negative influence affecting consumer attitude. Thus, online sellers should put more effort on minimising financial risk in order to influence positive consumer attitude to do online shopping with them. The perceived financial risk could be reduced by protecting customers' spending pattern and personal information, avoding misused of credit card details and eliminating overcharging price. In addition, product risk can be avoided by assuring the advertisers to sell quality products with clear pictures and description of the product. For non-delivery risk, it can be decreased by providing a place for online shopping customers to complain if they don't receive the delivery items on time and prompt action should be taken on it.

Secondly, convenience risk exerted positive effect on consumer attitude, which is consistent with the finding of Zhang, et.al [39]. The assumption made is that consumers may have a strong trust to the online seller site and therefore, they did not really concern on the issues of non-convenience aspect, such as handling of returned products and examining the quality of products featured in the site. The assumption is made based on the case of Kim and Prabhakar [41], who found that consumers' adoption of Internet banking is determined by a balance between trust and perceived risk.

Thirdly, this research has confirmed that positive affect of consumers' attitude on their online shopping behaviour. It shows that when a consumer formed a favourable attitude towards an object, for instance, positive attitude towards on line shopping, he or she tends to behave positively, for example, willingness to purchase online. This finding is consistent with the work of Chai and Pavlou [32], George [33], (Yang *et al.* [9] and Javadi, et.al [36]. Thus, individuals who have formed a favourable attitude towards online shopping are willing to commit in online purchase.

6.0 Conclusion

This study highlighted the common scenario in the study of consumers' perceived risk, attitude and online shopping behaviour. It seems that the negative effect of perceived risks on attitude of consumers is universal. In Malaysia online shopping context, it was confirmed that the negative effect of perceived risk influences the attitude of online shoppers. However, it was noted that not all risks, for example convenience risk, will negatively influence consumer attitude. It means that when consumers trust an online seller, they may accept issues related to return of purchased product, and they will execute some degree of tolerance in term of time taken to deliver the product. To validate the positive effect of convenience risk, it is suggested for future research to expand this study to cover more online sellers in Malaysia. Researchers may consider the inclusion of more dimensions of perceived risk, such as psychological, privacy and social risks, to examine whether or not they can influence attitude and online shopping behaviour of consumers.

References

[1] Lian, J. W., & Lin T. M. (2007). Effects of consumer characteristics on their acceptance of online shopping: Comparisons among different product types. Computers in Human Behavior, 24(1), 48–65.

[2] Li, N., & Zhang, P. (2002). Consumer online shopping attitudes and behavior: An assessment of research. *Information Systems proceedings of Eighth Americas Conference.*

[3] Bernama (2013). 2014 Budget: 25 mill internet users by 2015on. *New Straits Times.* Retrieved on 11 Nov 2013 at http://www.nst.com.my/latest/font-color-red-2014-budget-font-25-mill-internet-users-by-2015-1.384975

[4] Malaysia Digital Association (2012). Malaysia Website Ranking for February 2012. Retrieved on 15/3/2013 from http://www.digital.org.my/pdf/MDA-EM WebsiteRanking_Report_Feb12.pdf MCMC, (2012), p 19.

[5] Lobb, A. E., Mazzocchi, M., & Traill, W.B. (2006). Modelling risk perception and trust in food safety information within the theory of planned behaviour.

[6] O'Cass, A., & Fenech, T. (2003). Web retailing adoption: Exploring the nature of internet users web retailing behavior. Journal of Retailing and Consumer Services, 10, pp. 81- 94.

[7] Shih, H. P. (2004). An empirical study on predicting user acceptance of e-shopping on the Web. Information and Management, 41, pp. 351-368.

[8] Van der Heijden, H., Verhagen, T., Creemers, M. (2003). Understanding online purchase intentions: Contributions from technology and trust perspectives. European Journal of Information Systems, 12(1), pp. 41-48.

[9] Yang B, Lester D, James S.(2007) Attitudes toward buying online as predictors of shopping online for British and American respondents. Cyber Psychology & Behavior 2007; 10(2): 198-203

[10] Michelle Kovacs, Salomao Farias, Francisco Moura & Anderson Souza. (2011). Relationship Between Consumer Effort, Risk Reduction Strategies, and Satisfaction with the e- Commerce Buying Process: The Development of A Conceptual Formwork. International journal of management, 28: 316-329

[11] Lee, K.S., & Tan, S.J. (2003). E-retailing Versus Physical Retailing: A Theoretical Model and Empirical Test of Consumer Choice. Journal of Business Research, 56(11), 877–885.

[12] Ko, H., J. Jung, J. Kim, and S.W. Shim, (2010)"Cross-cultural differences in perceived risk of online shopping," Journal of Interactive Advertising, Vol. 4, No. 2: 20-29.

[13] Bauer, R. A. (1960). Consumer behavior as risk taking. In Cox, D. (Ed.), *Risk taking and information handling in consumer behavior* (pp. 389–398). Cambridge, MA: Harvard University Press

[14] Pires, G., Stanton, J. and Eckford, A., (2004) "Influences on the Perceived Risk of Purchasing Online," *Journal of Consumer Behaviour* Vol. 4, No. 2: 118-131.

[15] Mary Douglas and Aaron Wildasky (1982) http://www.svt.ntnu.no/psy/Torbjorn.Rundmo/Cultural_theory.pdf

[16] Ajzen, I., "the Theory of Planned Behavior, (1991) " *Organizational Behavior and Human Decision Processes* Vol. 50, 179-211.

[17] Pavlou, P. A., & Fygenson, M. (2006). Understanding and predicting electronic commerce adoption: An extension of the theory of planned Behavior. *MIS Quarterly, 30*(1), 115-143.

[18] Claudia, I. (2012). Perceived Risk when Buying Online: Evidence from a Semi-structured Interview. *Economics Series,* 22(2), 63-73.

[19] Kim, I.,(2010) "Consumers' rankings of risk reduction strategies in e-shopping," International Journal of Business Research, Vol. 10, No. 3: 143-148.

[20] Bhatnagar, A., Misra, S. and Rao, H. R., (2000)"on Risk, Convenience, and Internet Shopping Behavior," *Communications of the ACM* Vol. 43, No. 11: 98-105.

[21] Almousa, M. (2011). Perceived Risk in Apparel Online Shopping: A Multi-Dimensional Perspective. *Canadian Social Science,* Vol. 7, No.2, 23-31

[22] Saprikis V., Chouliara A., Vlachopoulou M., (2010) "Perceptions towards online shopping: Analyzing the Greek University students' attitude", *Communications of the IBIMA*, vol. 2010, pp. 1-13, 2010

[23] Chang, H., Chen, S. (2008), "The impact of online store environment cues on purchase intention: Trust and perceived risk as a mediator", *Online. Info. Rev.*, (6), 818 – 841.

[24] Dan, Y., Taihai, D., and Ruiming, L., (2007), "Study of Types, Resources and Their Influential Factors of Perceived Risk.

[25] Naiyi Ye, (2004) "Dimensions of consumer's perceived risk in online shopping", *Journal of Electronic Science and Technology of China*, vol. 2, no. 3, pp. 177-182.

[26] Chiu YB, Lin CP, Tang LL (2005). Gender differs: assessing a model of online purchase intentions in e-tail service. Int. J. service Industry Manag. 16 (5).

[27] Fishbein, M., & Ajzen, I. (1975). *Belief, Attitude, Intention and Behavior: An Introduction to Theory and Research*. Massachusetts, Addison-Wesley.

[28] Ajzen, I. and Fishbein, M., (1980) *Understanding Attitude and Predicting Social Behavior*. Prentice-Hall, Englewood Cliffs, NJ.

[29] Mitchell [29] (1999) Mitchell, V. W. (1999). Consumer perceived risk: conceptualization and models. *European Journal of marketing*, 33(1/2), 163-195.

[30] Barnes, S., Bauer, H., Neumann, M. and Huber, F. (2007), "Segmenting cyberspace: a customer typology for the internet", *European Journal of Marketing*, **41**(1/2), 71-93.

[31] Pavlou, P.A. (2003). Consumer intentions to adopt electronic commerce – incorporating trust and risk in the technology acceptance model. *International Journal of Electronic Commerce, 7*(3), 101–134.

[32] L.Chai and P. Pavlou, (2004) "From Ancient to Modern: A Cross-Cultural Investigation of Electronic Commerce Adoption in Greece and the United States " *Journal of Enterprise Information Management,* 17, 6.

[33] George, J. F. (2004). The theory of planned behavior and Internet purchasing. *Journal of Internet Research, 14*(3), 198

[34] Swinyard, W. R., & Smith, S. M. (2003). Why People Don't Shop Online: A Lifestyle Study of the Internet Consumers. *Psychology and Marketing, 20*(7), 567-597.

[35] Forsythe, S., Liu, C. L., Shannon, D., & Gardner, L. C. (2006). Development of a scale to measure the perceived benefits and risks of online shopping. Journal of Interactive Marketing, 20(2), 55-75

[36] Javadi , M., Dolatabadi, H.. Nourbakhsh, M. Poursaeedi, A., & Asadollahi, A. (2012) , "An Analysis of Factors Affecting on Online Shopping Behaviour of consumers", *International Journal of Marketing Studies*; **4**(5), 81-98.

[37] Karayanni, D. A. (2003). Web-shoppers and non-shoppers: Compatibility, relative advantage and demographics. *European Business Review, 15*(3), 141-152.

[38] Hair, J. F., R.E. Anderson, R.L. Tatham, and W.C. Black, —Multivariate Data Analysis, Fifth Edition, NY, Upper Saddle River: Prentice-Hall: 1998.

[39] Zhang, L., Tan, W., Xu, Y. & Tan, G. (2012) , "Dimensions of Consumers' Perceived Risk and Their Influences on Online Consumers' Purchasing Behaviour", *Communications in Information Science and Management Engineering*, **2**(7), 8-14.

[40] Masoud, E.Y (2013). The Effect of Perceived Risk on Online Shopping in Jordan. *European Journal of Business and Management,* Vol.5, No. 6, 76-87.

[41] Zakuan, N and Saman, M.Z.M, (2009). "Lean Manufacturing Concept: The Main factor in Improving Manufacturing Performance", International Journal of Manufacturing Technology and Management, 17(4) pp. 353-363

Experimental testing of centrifugal pump: small and medium sized enterprise product

R Ismail, F Paddiyatu, M Khafidh, S Nugroho, S Sugiyanto and J Jamari
Department of Mechanical Engineering, Diponegoro University
Jl. Prof. Sudharto Kampus UNDIP Tembalang Semarang 51275 Indonesia

E-mail: r_ismail@undip.ac.id, ismail.rifky@gmail.com

Abstract. This paper reports the experimental testing for centrifugal pump for fisherman ship, manufactured by small and medium sized enterprises in Central Java Province, Indonesia. The research covers material analysis, component observation, endurance and vibration test. Six centrifugal pumps are tested and three main pump components are discussed: shaft, bearings and seals. The results show that the material of the shaft is predicted to support and transmit the load from the engine to impeller. The problem found in the tolerance and geometry accuracy of the shaft which causes difficulties during assembling process, excessive wear and leakage during testing. From the endurance and vibration test, the ball bearings fail and lock the shaft due to the fatigue on the rolling elements and raceways. The oil seal and water seal also fail in maintaining the oil and water in the chamber and induce the unlubricated system for the ball bearings. Some suggestions are delivered to improve the product quality of the centrifugal pump. A good quality of the centrifugal pump for fishermen ship and long life span is expected to be produced by local SMEs to win the free trade competition in the Indonesian market.

1. Introduction

In the last two decades, Small and Medium Sized Enterprises (SMEs) in Indonesia have been considered as pillar of Indonesian economy. In 2011, Minister of Cooperatives and SME, Republic of Indonesia stated in his annual report that 98% of all Indonesian enterprises (around 55 million companies) are categorized as SME and employ more than 90 million employees which equal to more than 95% of Indonesian employees [1]. SMEs also have historically proven as the significant players in the Indonesian economy and survived in several global and regional economic crises. Nowadays, big challenge comes to SMEs in Indonesia due to regional free trade agreement among South East Asia countries and China, as a consequence of global market. Indonesian trade policy is changed from a protected to an open economy due to the implementation of the regional free trade agreement. This situation could be a threat for the impressing and positive growth of SMEs during a last decades. The impact of this policy could be received by SMEs in Central Java Province Indonesia, especially metal manufacturer SMEs. They will found that the competition in selling the metal product in the market is not easy anymore due to global competition.

In the future, there are pessimistic opinions that Indonesian SMEs will not survive in the era of trade liberalization, although some reports assured that until 2011, no significant impact of free trade agreement on SMEs sustainability [2]. However, based on the field observation in Central Java

Province, metal manufacturer SMEs are facing various constraints, including lack of technology application, product development, qualified human resource and product quality assurance [3]. These constraints are negative factors for SMEs in confronting the regional free trade agreement. This leads researchers conduct evaluation and testing on the metal product of SMEs and propose some improvements. One of the projects is supporting and evaluating SMEs which manufacture metal product for marine application. These projects are proposed to evaluate the technology application, design, material and component selection, manufacturing process, product performance and quality assurance.

One of the products which should be evaluated is centrifugal pump for marine application, manufactured by several SMEs in Central Java Province, Indonesia. Figure 1 depicts the installation of the centrifugal pump in the fisherman marine boat, connected to the marine engine using a belt to transmit the power. Figure 2 shows the bearings and shaft in the centrifugal pump.

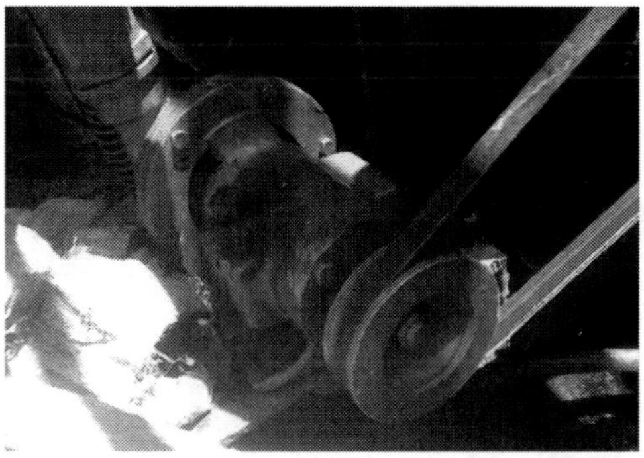

Figure 1. The centrifugal pump, manufactured by SME: the installation of the pump on the fisherman boat.

Figure 2. Bearings and shaft of the pump.

The research on centrifugal pump of SME focuses on two steps: product evaluation and product improvement. The product evaluation step investigates material and components selection and defines problems of centrifugal pump during the tests. Experimental evaluation and condition monitoring are frequently used by researchers to diagnose and investigate the condition of existing pump components [4-7]. The improvement step focuses on the suggestions to improve the quality, endurance and life time of the centrifugal pump. The present work focuses on the first step of the research (product evaluation) and discusses several suggestions for SMEs in Central Java Province which produce centrifugal pump. This paper reports the experimental tests for the centrifugal pump, manufactured by SMEs in Central Java Province which covers: material and component assessment, endurance test and vibration test and suggests improvements to increase the quality and life time of the pump.

2. Research Methodology

Product evaluation is initiated from components and material selection, manufacturing and assembling process observation, quality control assessment and endurance performance measurement. Six centrifugal pumps are tested in Laboratory for Engineering Design and Tribology, Diponegoro University. One of the pumps is disassembled and the components are tested to observe its properties and functions whereas five pumps are tested in test rig. A disassembled pump is observed for the three important components: shaft, bearings and seals. Based on the field observations and the interviews with the fishermen, it was reported that the failure of the SME pumps is related with these components. The chemical composition test, microstructure observation, hardness test and tensile test are conducted to the pump shaft material. A series of geometry test is carried out to observe the tolerance and geometry accuracy of the shaft. For the bearing, a SEM observation is conducted to study the new and used bearings. The new and used component observation is also conducted to the pump seals.

Five centrifugal pumps, manufactured by SMEs in Central Java, are prepared to be evaluated in the test rig, depicted in Figure 3. Two imported products from foreign competitor are involved in this research to compare the components selection and the product performance. These pumps are tested in a rig to evaluate the endurance of the pumps. Figure 3 depicts a testing equipment where each pump are connected to a 1 HP electric motors (three phases) to recirculate the sea water. The pumps recirculate the sea water for 12 hour non-stop for a day with 2200 rpm and it is repeated until 40 days. A vibration test, showed in Figure 4, is carried out for the new pumps and for the pumps with the high noise due to initial failure of the bearings or shaft. The pump is moved to the vibration test when high noise is found in the endurance test rig.

Figure 3. A test rig to evaluate the performance of centrifugal pumps.

Figure 4. Vibration testing of centrifugal pump and the two sensors on the bearing.

3. Results

3.1. Shaft

The result from the spectrometer test for determining the chemical composition of the shafts is listed in Table 1. Only the main elements are reported in this table whilst the other is neglected due to the small contributions. Based on the laboratory test, hardness number of the shaft is 30 HRC whereas the tensile strength is 808 MPa. The material of the pump shaft, made from stainless steel with high tensile strength and hardness, is predicted to support and transmit the load from the engine to the impeller. Some comparisons of the SME shaft with the imported pump shaft product were discussed in the previous paper [8]. The microstructure of the shaft below microscope observation is shown in Figure 5 (a) with 100 x magnifications and Fig. 4 (b) 500 x magnifications.

Table 1. Chemical composition of the stainless steel shaft.

Chemical Composition	**Fe**	**C**	**Si**	**Mn**	**Cr**	**Ni**	**Cu**
%	71.2	0.124	0.340	10.8	14.6	1.23	1.35

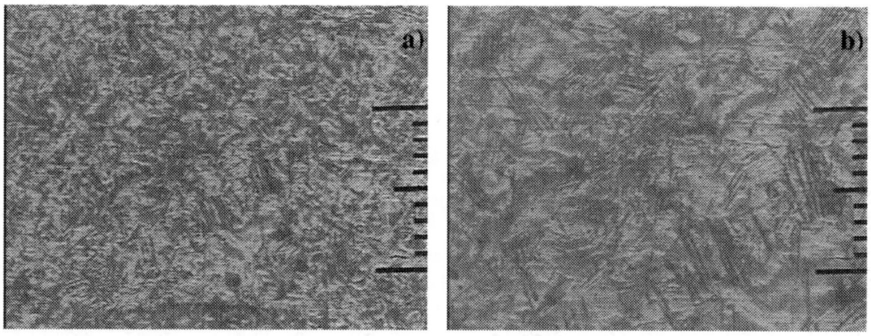

Figure 5. The microstructure of the centrifugal pump shaft: (a) 100 x magnifications and (b) 500 x magnifications

However the problem is found for the shaft when a series of geometry test is carried out to 20 pump shafts for investigating the tolerance and geometry accuracy. Two critical positions on the shaft where the bearings are located are highly varied between (-) 0.3 mm until (+) 0.3 mm for 20 mm of the shaft diameter. The turning operator pays less attention on the accuracy of the shaft diameter and causes the undesirable tolerance and fits between the shaft and bearings and leads the excessive wear on the shaft (for the large clearance fit) and the damage of the bearing during assembly process (for the shrink fit). The normal clearance should be 0.05 – 0.1 mm.

3.2. Bearings

There are 4 ball bearings in a centrifugal pump shaft to support the radial load, see Figure 2. The bearings are observed in two conditions: the new bearing and the failure bearing after the endurance test. From the endurance test it is found that 1 pump failures on the 21st day of the test. The shaft is locked due to the failure of the bearings. The other 4 pumps reach the last day (40th day) in the endurance test with relatively high noises and vibrations, indicating an initial failure of the bearings. In a centrifugal pump, defective bearing can cause more number of serious problems such as abnormal noise, leakage, high vibration, etc. [5]. The results of the vibration test for the new pump and the nearly failure pump are depicted in Figure 6 (a-b) and Fig. 7 (a-b), respectively. These figures were obtained from the vertical vibration sensor. The amplitude versus time and the FFT graphics are presented as the common vibration analysis [6].

The vibration monitoring records the range of amplitude of the failure pump on the 21st day (Figure 7a) is 50 times higher than the range of amplitude of the new pump (Figure 6 a). It indicates that the failure of bearing produce higher vibration and noise. After the pump fails and the bearings are locked then the pump is disassembled and the bearings are checked and cut to observe the condition of the inner raceway, outer raceway and the spherical balls. SEM analyses for the failure bearings are depicted in Figure 8-10 (a-b). Figure 8 shows the fatigue failure on the rolling element bearing where

IOP Conf. Series: Materials Science and Engineering **58** (2014) 012008 doi:10.1088/1757-899X/58/1/012008

spalls are found on the spherical ball surface. Figure 9 and 10 depict the wear on the inner and outer raceway of the ball bearing. The failure mode limiting bearing life is mostly rolling-element fatigue of either a bearing race or rolling elements. Rolling-element fatigue is extremely variable but is statistically predictable depending on the steel type, steel processing, heat treatment, bearing manufacturing and type, lubricant used and operating conditions [7]. In the case of the centrifugal pumps product there are some causes of the bearing failure: (i) SME uses a low quality bearings, (ii) the leakage of the lubricant in lubricant chamber leads the unlubricated system and (iii) the tight fits due to excessive geometry tolerance of the diameter shaft cause the ball bearing received higher force during assembling process.

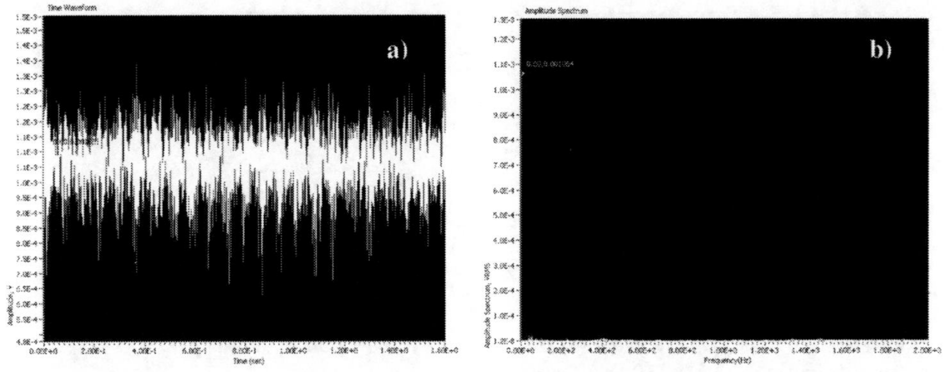

Figure 6. The vibration monitoring on vertical direction of the bearing for the new pump: (a) amplitude vs time graphic and (b) FFT graphic.

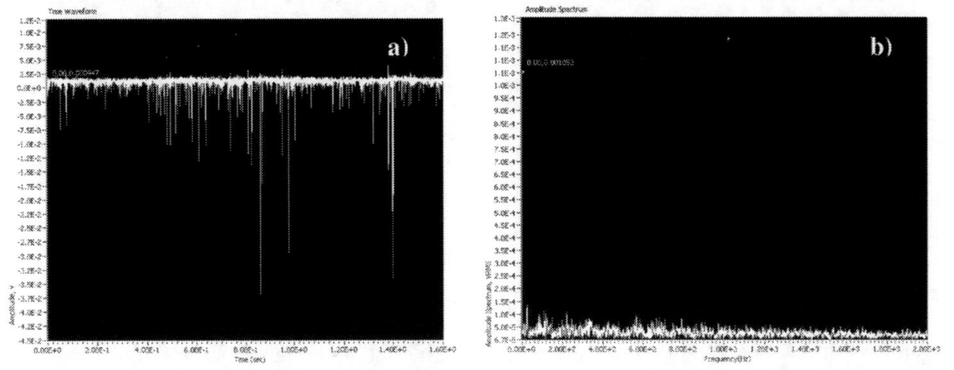

Figure 7. The vibration monitoring on vertical direction of the bearing in the nearly failure pump: (a) amplitude vs time graphic and (b) FFT graphic.

3.3. Seals

There are two types of seal in the centrifugal pumps: oil seal to avoid the lubricant leakage and gasket to avoid the water leakage, see Figure 11. SME uses a low quality of oil seal and manufacture the gasket by them self from the non-standard rubber. The result can be predicted easily, the leakages are found. During the endurance test, all of the pumps indicate the initial water leakage after 3-5th day of running and the oil leakage is identified after 8-10th day of running.

Figure 12 depicts the oil leakage on the pump holder in the test rig which causes dry contact situation on the ball bearings. It can be caused by: (i) a low quality of the oil seal where the springs easy to break and induce the leakage of the lubricant, see Figure 13, (ii) the failure of the seal because of assembling process, (iii) a large clearance between the shaft and oil seal and (iv) the sharp shape of the shaft break the oil seal when it is pressed into the shaft.

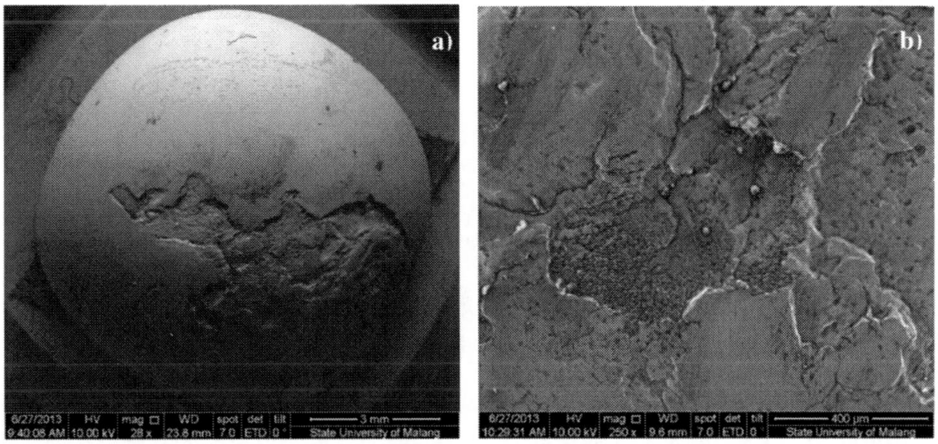

Figure 8. SEM on the ball bearing: (a) failure on the ball and (b) zoom view of the failure ball.

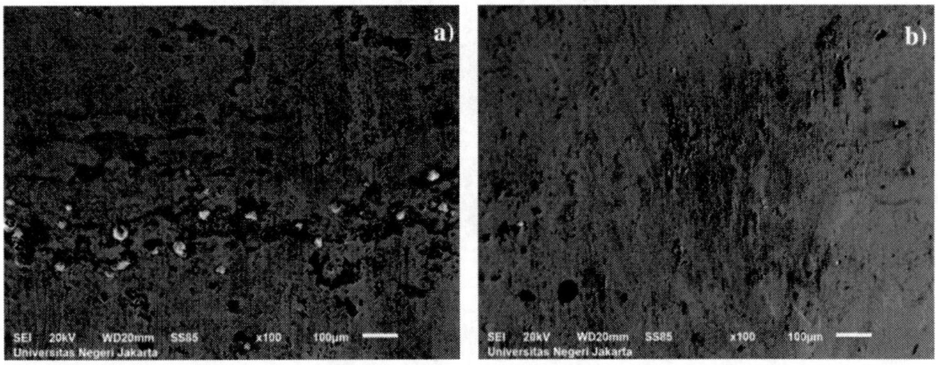

Figure 9. SEM on the inner raceway of the failure bearing. Pitting wear is found at (a) the centre of the inner raceway surface and (b) several distances from the centre.

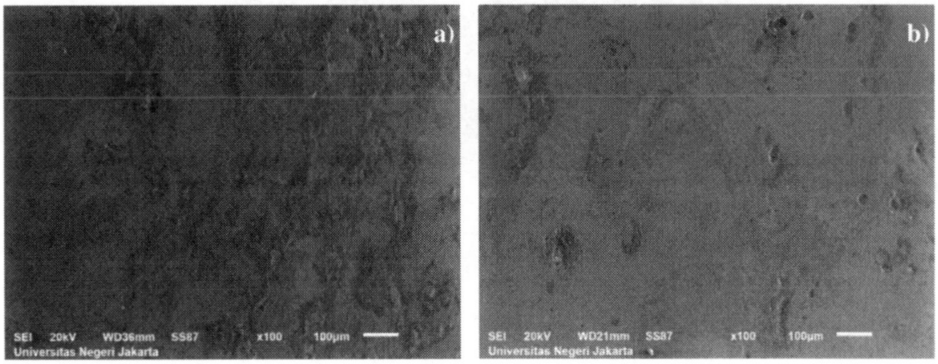

Figure 10. SEM on the outer raceway of the failure bearing. Fatigue wear is found at several points at the outer raceway surface (a-b).

IOP Conf. Series: Materials Science and Engineering 58 (2014) 012008 doi:10.1088/1757-899X/58/1/012008

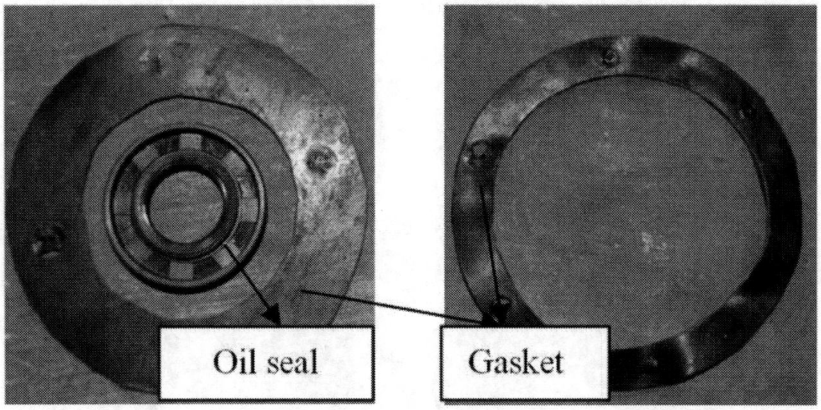

Figure 11. Oil seal and water seal (gasket) in centrifugal pump.

Figure 12. The leakage of the lubricant during the endurance test.

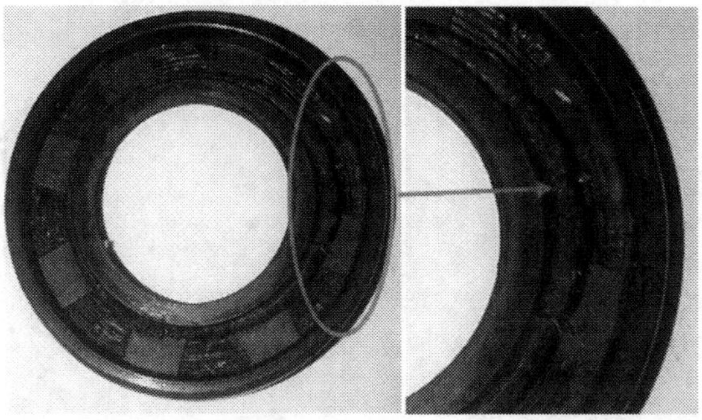

Figure 13. The failure of springs of the oil seal after the pump is disassembled.

4. Suggestion for Improvement

In this section some suggestions are given to SMEs for improving the quality, endurance and life time of the centrifugal pump. The suggestions are listed in Table 2 which covers the shaft, bearing, seal and quality control. The improvement of the shaft is related to the dimension accuracy, chamfer on the seal entrance and the quality control. The chamfer around 2-3 mm is suggested to avoid the seal crack during assembly process. The quality control, such as the use of go-no-go gauge, should be applied to increase the dimension accuracy of the shaft.

Table 2. Suggestions for SMEs to improve the centrifugal pump quality.

Target	**Improvements**
Shaft	▪ Increase the accuracy of the dimension to obtain the best fit between the shaft and bearings ▪ Create chamfer on the entrance of the oil seal ▪ Quality control for the shaft dimension
Bearing	▪ Use good quality bearings ▪ Assembly the bearings properly using the new press equipment ▪ Ensuring the lubricated system on the bearings
Seal	▪ Use good quality oil seal and sea water seal ▪ Assembly the seals properly to avoid seal cracking ▪ Ensuring the lubricated system on the bearings
Quality control	▪ Quality control should be conducted for every pump components: shaft, impeller, housing, pulley, etc. ▪ SMEs should provide an endurance test rig to check some sampling of the centrifugal pumps ▪ A running-in stage of centrifugal pump is recommended after the assembly process and before the pump is wrapped

The suggestion for the bearing covers the selection of good quality bearings, the new press equipment to assembly the bearings and ensuring the lubrication system on the bearings. The new press equipment, see Figure 14, is suggested to avoid the bearing failure during assembly process. The suggestion for the seal focuses on the selection of good quality seals and the assembly process of the seal.

A last suggestion is related to the quality control and assurance. Previously, SMEs did not provide controller for checking the components dimension and quality. The quality control officer checks all components, manufactured locally by SMEs and manufactured by third parties, to obtain a proper dimensional and tolerance accuracy. An endurance test rig and running-in test rig are also suggested to the SMEs to assure the quality of the centrifugal pumps. The running-in stage of the pumps is required for ensuring the conforming contact and the fully lubrication system before the pump is installed to the ship and receive the high load. Every contacting component, such as bearing, gears, camshaft, piston, etc. is suggested to experience the running-in phase with lower load before a real maximum load is applied to the components. The running-in stage of the pump is expected to be finish in few hours and suggested to be conducted by SMEs before the pump is wrapped and delivered to the customer.

The proposed improvements are still adapted and applied by SMEs in their manufacturing process. Some centrifugal pumps are manufactured with researcher supervisions. These improvements increase the centrifugal pumps around 10-20% but the productive time of the pump is predicted to increase 3-4 times. The results of the improvement are still tested in the endurance test rig with the good preliminary results of the noise and vibration.

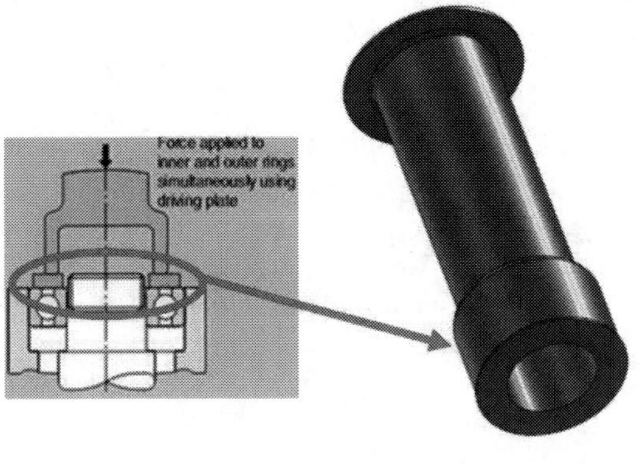

Figure 14. A new design for pressing tool to assembly the bearings into the shaft and the pump housing based on the literature is proposed to SMEs.

5. Conclusion

The paper reported the evaluations and assessments of the centrifugal pumps, manufactured by SME in Central Java Province Indonesia. The evaluation and assessment are required to improve the quality of the centrifugal pumps due to a higher competition on the marine product after the implementation of regional free trade agreement in South East Asia region. The evaluation step covers the shaft, bearing and seal components as well as the quality control and product assurance during the manufacturing and assembly process in SMEs. The results show that the pump shaft, made from stainless steel with high tensile strength and hardness, is able to support and transmit the load from the engine to the impeller. However the problems of the shaft are the tolerance and geometry accuracy which lead the excessive wear on the shaft (for the large clearance fit) and the damage of the bearing during assembly process (for the shrink fit). The bearings of a pump are failure on the 21st day of the endurance test and lock the shaft. SEM analyses for the failure bearings are observed to determine the fatigue failure on the rolling element and raceway. During the endurance test, all of the pumps indicate the sea water leakage after 5th day of running and the oil leakage is identified after 10th day of running which indicates the problem of the quality of seal and assembling processes. Some suggestions have been delivered to improve the product quality and assurance of the centrifugal pump. The good quality centrifugal pump with a product assurance and long life span is expected to be produced by local SMEs to win the free trade competition in the Indonesian market.

6. Aknowledgement

The authors acknowledge to Directorate General of Higher Education, Ministry of Education and Culture, Republic of Indonesia for MP3EI Research Grant with contract No. 545/SK/UN7.3.3/VI/2013 to conduct this researches.

7. References

[1] Hasan S, Awakening of Cooperative and Small and Medium Sized Enterprises toward People Prosperity, Annual Performance Report of Ministry of Cooperation and SME, Republic of Indonesia, Jakarta, Indonesia, 2011.

[2] Tambunan T, The Impacts of trade liberalization on Indonesian small and medium-sized enterprises, TKN Policy Paper, International Institute for Sustainable Development (IISD), Canada 2011.

[3] Lestari H, Ismail R and Mansur A 2014 *App. Mech. Mat.* **493** 486

[4] Schaefer G 1999 *World Pumps* **1999** 34
[5] Sakthivel N R, Sugumaran V and Babudevasenapati S 2010 *Expert Sys. Apps.* **37** 4040
[6] Muralidharan V and Sugumaran V 2013 Measurement **46** 3057
[7] Beebe R S 2004 *Predictive Maintenance of Pumps Using Condition Monitoring* (Amsterdam: Elsevier)
[8] Ismail R, Sugiyanto S, Tauviqirrahman M and Jamari J 2012 *Proc. Int. Conf. on Chemical and Material Engineering in Diponegoro University* (Semarang: Central Java - Indonesia) MSD 05

Investigation on mechanical properties of AlZrCr- Al_2O_3 nanocomposites fabricated by stir casting

Kirman M.[1a,2], Anne Zulfia[1b], Sutopo[1c] and Bambang Suharno[1d]

[1]Faculty of Engineering, University of Indonesia, Depok 16424, Indonesia
[2]Center of Material Technology, Agency of the Assessment and Application of Technology, Building 224, Science and Technology Park (Puspiptek), South Tangerang 15314, Indonesia

E-mail: [a] kirmanmt@yahoo.com, [b] anne@metal.ui.id, [c] sutopo@metal.ui.ac.id, [d]suharno@yahoo.com

Abstract. Aluminum alloy composite with nano Al_2O_3 reinforcement will be designed to have good mechanical properties that correspond to its application. The addition of nano Al_2O_3 in aluminum is to increase strength and stiffness. In this study aluminum matrix made as a master alloy with the addition of zirconium (Zr) and cerium (Ce) to form Al-Zr-Ce alloys which was reinforced with nanoscale alumina particles known as nano aluminum composite which has high strength and stiffness. Master alloy Al-Zr-Ce used as a matrix content of 0.12 wt% Zr and 0.13 wt% Ce, while Al_2O_3 nano particles (<100nm) used as reinforcement was various in the range of 0-3 Vf%. Aluminum composite was produced by stirring of molten metal with a rotational speed of 500 rpm at a temperature of 750°C in an inert argon gas environment then characterized both mechanical properties and microstructure analysis. The tensile strength increased with increasing Al_2O_3 nano particles up to 1 Vf%. Mechanical properties of composites were slightly increased and there was no significant change in elongation and hardness, perhaps due to the non-uniformity distribution or clustering formation of particles in the matrix.

1. Introduction

Researchers have conducted numerous studies to improve the mechanical properties of aluminum. Improved mechanical properties of aluminum can be done by adding alloying elements into aluminum or can be done by adding a reinforcing material in aluminum as Al_2O_3, SiC, CNT etc. Several studies have reported that the addition of rare earth metals to the aluminum can increase the tensile strength and electrical conductivity. If the aluminum containing impurity elements such as Fe and Si added rare earth metals such as Ce then some of these elements will be dissolved in the crystalline aluminum and others will form a metastable compound or the second phase as $FeAl_3$ or Ce_3Si_2 [1]. Pengfei Li et al[2] also reported that the addition of rare earth metals can alter the distribution of impurities in aluminum. Rare earth metals to form intermetallic bond with impurities and they will segregate at the grain boundaries. The content of rare earth metals on aluminum will refine the grain structure of aluminum so it will improve the mechanical properties. Similarly, the addition of Zr element in aluminum will also improve the mechanical properties. They reported that the addition of 15 % Zr in aluminum will result in a solution of a second phase or a metastable compound Al_3Zr which has a very small grain size.

Commercially pure aluminum has been used as an electrical transmission lines for decades, because it is cheaper than copper. The use aluminum as a conductor tends to increase. Now most of the electrical transmission network has use commercially pure aluminum conductor. Nevertheless, the uses of aluminum as an electrical conductor still need to optimize the required properties as a conductor wire. This has prompted many researchers conduct research to improve the properties of the electrical conductivity, heat resistance and mechanical properties. Akhmad Zamroni[3] has do the addition of 0.1% - 0.3 wt% cerium zirconium in pure aluminum. The addition of rare earth metals cerium at 0.05 wt% and 0.72 wt% can increase electrical conductivity. The addition of zirconium increase durability and lower heat conductivity but cerium additions will increase the electrical conductivity. A similar study conducted by Zulkifli[4] by adding 1.5 % - 4.0 wt% Zr and Gunawan[1] by adding Zr 0.04 wt% and 0.13 wt% lanthanum.

Other efforts are needed to improve the behavior of the aluminum conductor is to improve the mechanical properties. One effective way to improve the mechanical strength of the material is the use of nanoparticles as reinforcement in aluminum matrix. It has been well understood that the presence of a dispersion of fine particles (<100 nm) in a metal matrix greatly improves creep resistance, though with only a small volume fraction (<1%)[5]. Research the addition of nanoparticles in aluminum alloys has been performed several people, among others Takashi et al[6] who reported the effect of adding nanoparticles of titanium-zirconium in aluminum alloy. To increase knowledge in the field of strengthening of aluminum alloy, this study was done different things, namely the strengthening of aluminum alloy Al-Zr-Ce by nanosize particles with 0.2-3% olume fraction of Al_2O_3.

2. Experimen setup and Procedure

2.1. Starting materials

The material used were pure aluminum, Al-Zr5% and Al-Zr10% were supplied by Hunan Jinlianxing Special Material Co.Ltd. as a master Al-Zr5% and Al-Zr10%. These materials were manufactured as alloy of matrix composite in the form Al-Zr-Ce. The particles of Al_2O_3 was <80 nm as a reinforcement of composite which was supplied by US Research Nanomaterials, Inc. Houston-USA.

The chemical composition of composite master alloy Al-Zr-Ce conducted by Field Emission - Scanning Electron Microscope (FE-SEM) under Energy Dispersive Spectrometer (EDS) and the results of Al-Zr-Ce was shown in the table 1.

Table 1. Chemical composition of Al-Zr-Ce alloy

Alloy material	Weight %					
	Al	Zr	Ce	Si	Fe	others
Al-Zr-Ce	98.30	0.12	0.13	0.39	0.88	0.18

2.2. Production of Al-Zr-Ce/$Al_2O_{3\,np}$ nano composites

Al-Zr-Ce/ Al_2O_{3np} nano composite was produced by stir casting method as is shown in figure 1. The reinforcement of Al_2O_{3np} was varied from 0 to 3 Vf%. The casting process start from feeding reinforcements into the matrix by inserting the particle into aluminum foil capsule before put it in the middle of stack small block of aluminum alloy in the furnace. Melts aluminum was degassed by inert argon gas and it were stirred with a rotational speed of 500 rpm for 2 to 3 minutes. When nanosize Al_2O_3 particles were added in the molten Al alloy, the viscosity of the molten Al alloy significantly increased. Thus, a higher melt temperature of 750°C was used to ensure the flow ability inside a mold. Then the melt metal matrix composite poured into a permanent tensile test mold (figure 2). After pouring is over, the melt was allowed to cool and solidify in the mold.

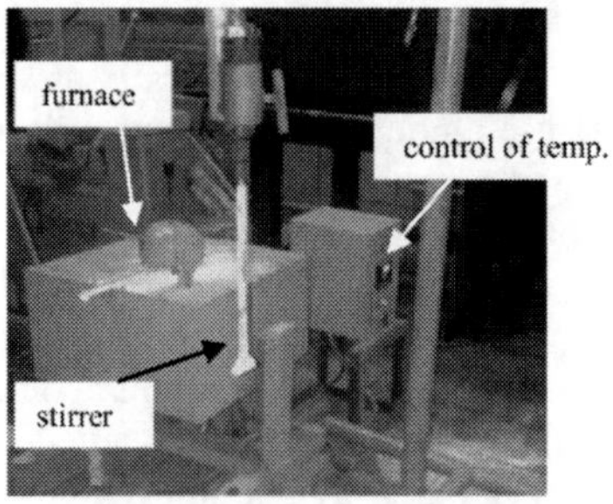

Figure 1. The experimental apparatus for producing the composites by stir casting method

2.3. Mechanical Tests

Tensile specimens prepared three pieces from each composition as shown in Figure 3. The dimensions of tensile test specimens were gauge-length 32 mm and diameter 6 mm. The tensile properties of the nanocomposite specimens were tested at universal tensile testing machine Carl Schenck RME100 in according with the ASTM E8 standard. The hardness of the samples was measured using macro hardness Brinnel by applying load of 31.25kg with 2.5mm diameter of indenter and the load was applied for 20 seconds. In order to eliminate possible nonhomogenized material effect, a minimum of five hardness readings were taken for each specimen at different locations of the test samples. The specimens for hardness were prepared with dimension of sample is 50 mm length and 10mm thickness. First the surface was ground with emery paper to remove the oxide layer and cleaned. On the other hand, microstructural analysis was prepared by metallography procedure, start from grinding the specimens of 10x10x10 mm through 80, 120, 240, 500, 700, 1000, 1200 grit SiC papers followed by polishing with 6, 3, 1, ¼μm diamond pasta and etching with Keller's Reagent. The microstructures were obtained by viewing the samples at different magnification levels on optical microscope and FE-SEM.

Figure 2. Tensile specimen permanent mold[7]

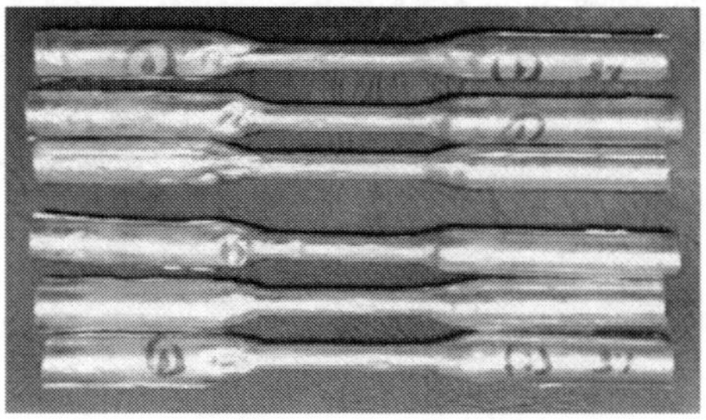

Figure 3. Tensile test specimens

3. Results and discussion

3.1 Microstructur of Al-Zr-Ce /Al_2O_{3np} composite

Aluminum alloy Al-Zr-Ce as a matrix of composite has been examined using an optical microscope and FE-SEM images as shown in figure 4. Microstructure of Al-Zr-Ce shows a fine grain size of aluminum alloy. Zirconium was dissolved in pure aluminum matrix whereas cerium went to the grain boundaries together with other impieties such as Fe and Si. The fine grain size is similar to what was reported by AA Rao[8] that adding 0.2 wt% Zr to 99.7% pure Al which causes a fine grain size.

Figure 5 shows microstructures of the aluminum matrix composite with 0.2 Vf% nanoparticle Al_2O_3 content. After addition nanoparticles to molten matrix, the grain refinery has occurred on the matrix even though only occurred in several areas. The occurrence of uneven grain refinement is expected as a result of non-homogeneous distribution of cerium. Distribution of cerium in the composite can be seen in Figure 6 where the elements cerium is marked by a bright white color phase.

Microstructure of Al-Zr-Ce unreinforced has small/fine grain and same precipitate in grain boundary. It is assumed that there is a primary Al in the material which was investigated at low magnification but at higher magnification under EDS, all of phase formed was identified contain of Al, Si, Zr, Ce and Fe in the table 2. The Chinese script was also seen the matrix.

Table 2. Chemical composition of Al-Zr-Ce /Al_2O_{3np} composite by EDS in figure 4

Spectrums	Weight %				
	Al	Zr	Ce	Fe	Si
1	99.87	0.13	-	-	-
2	71.00	-	20.23	-	8.77
3	95.09	-	-	4.91	-
4	93.64	-	0.31	6.05	-

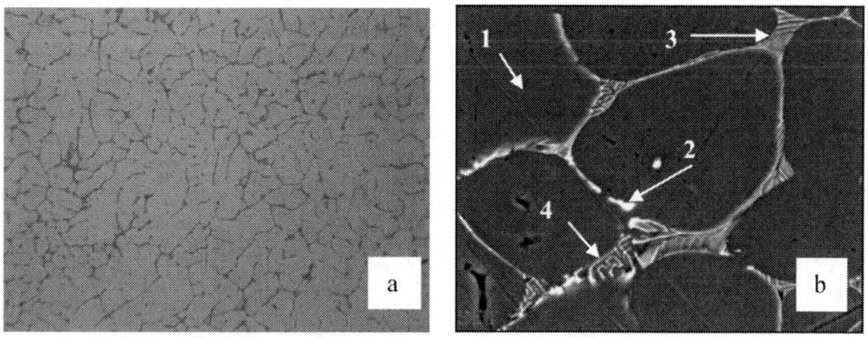

Figure 4 Microstructures of the as-cast Al-Zr-Ce alloy without reinforcement, (a) optical microscopy image (magnification with an uniform fine grain size and (b) FE-SEM image, EDS analysis Zr in the grain and Ce (bright white color) and impurities Fe and Si at grain boundary, the microstructure of matrix observed as Chinese script.

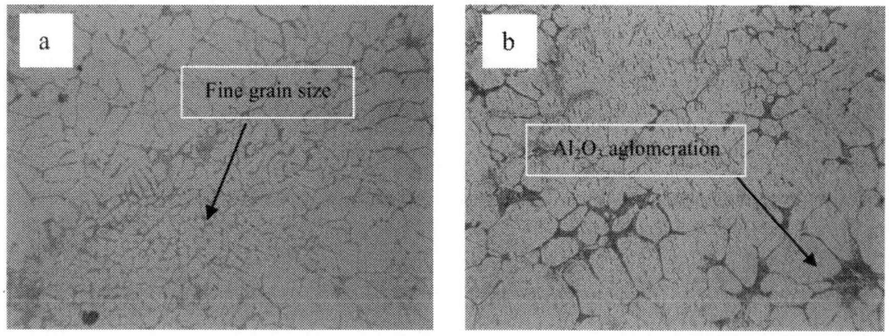

Figure 5. A typical microstructure of Al-Zr-Ce composite, (a) finer grain size of the metal matrix Al-Zr-Ce in local area then whole area and (b) microstructure of composite with nanoparticle Al_2O_3 agglomerations

Another observation from the FE-SEM test result in figure 6 showed clearly that there is a black dot on the microstructure of composite which identified as an agglomeration of Al_2O_3. Analysis Al_2O_3 particles form clusters with a diameter ± 5μm. Effect of clustering particle will cause Orowan's strengthening mechanism in composites cannot work well because the particles do not inhibit intergranular dislocations due to a larger particle size of 100nm[9]. Some porous was observed at interface which was indicated as dewetting.

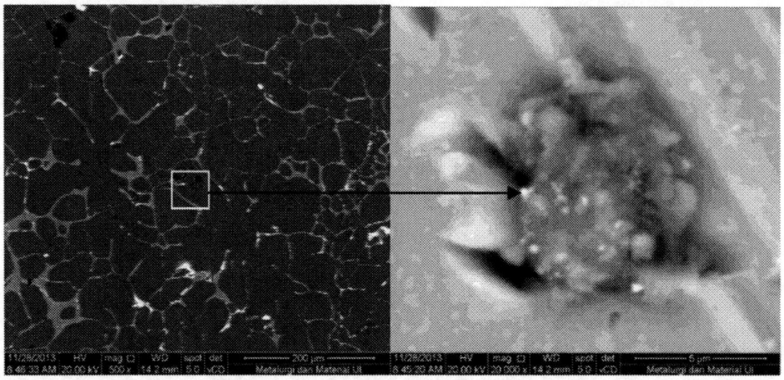

Figure 6 Microstructure image by SEM of Al-Zr-Ce /1%Al_2O_3 composite with Vf 1% nanoscale alumina, the cluster Al_2O_3 particles is clearly appear in magnification 20,000 times.

3.2 Mechanical properties of Al-Zr-Ce /Al_2O_{3np} composite

Al-Zr-Ce alloy nano composites with the addition of reinforcing particles 0 to 3.0 Vf% Al_2O_3 have evaluated ini the mechanical behavior. The mechanical properties of the composite Al-Zr-Ce /Al_2O_{3np} with different tensile strength, elongation and hardness composites are presented in Figure 7. The tensile strength increases with an increase in the amount of Al_2O_3 nanoparticles up to 1 % and then decreased when volume fraction of Al_2O_3 nanoparticles more than 1 % Vf. The slight increase in tensile strength is expected due to the effect of smoothing the grain or the effect of Al_2O_3 nanoparticles. Contrary, the tensile strength decreases when the nanoparticles Al_2O_3 more than 1 % Vf due to the effects of the more particle agglomeration. This result is decline from the study reported by Pradeep Rohatgi[10] that nanostructure aluminum can be 10 times stronger than conventional aluminum alloys . Failure of an increase in tensile strength up to 10 times as a result of porosity and nanoscale alumina particles agglomerations. This happens because the smaller the particle, the more easy-going particle clustering and distribution methods of nanoparticles with a mechanical stirring system with 3 minutes stirring time is not optimal to distribute the particles evenly. And also, argon gas was not used to protect the atmospheric air during the strring process that allows the formation of porosity in the composite.

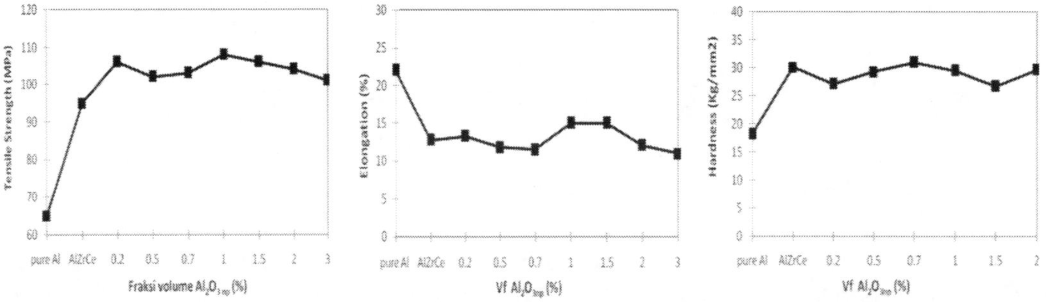

(a) (b) (c)

Figure 7 Mechanical properties of Al-Zr-Ce /$Al_2O_{3\ np}$ composites, (a) tensile strength, (b) elongation, (c) hardness

The addition of the aluminum alloy Al_2O_{3np} Al-Zr-Ce for elongation and hardness do not change much. Increasing the elongation at Vf 1 and 1.5% in Figure 7 (b) is assumed to be the effect of grain size on the homogeneous matrix in which there was not found a finer grain size than the grain size for the whole area.

4. Conclusions

Hardness and tensile strength of aluminum alloy (Al-Zr-Ce) that reinforced with various percentage of nanoparticles Al_2O_3 (0-3 Vf%) was examined and compared with matrix alloy. Results of the analysis indicate that by increasing number of nanoparticles, tensile strength of the composites was slightly increased by adding of Al_2O_{3np} up to 1 Vf% whereas ductility and hardness are relatively no change by adding of Al_2O_{3np}. Observation of microstructure by FE-SEM, and optical microscope also be identified that distribution of nanoparticle on the aluminum matrix are very poor and the grain size of aluminum matrix of composite have changed to be fine grain in localized area of the metal matrix.

Acknowledgements

Authors would like to thanks to Higher Education, Ministry of National Education, Republic of Indonesia for financial support under Hibah Desentralisasi PUPT, 2013 with contract no. 2385/H2.R12/HKP.0500/2013

References

[1] Gunawan 2000 Effect of Zirconium and Lanthanum Addition on Electrical Conductivity and Heat Resistance of Aluminum Conductor Wire, *Thesis S2 Material Science, University Indonesia*, 2000

[2] Li Pengfei, Wu Zhigang, Wang Yunli, Gao Xizhu, Wang Zaiyun, Li Zhiqiang 2006 Effect of cerium on mechanical performance and electrical conductivity of aluminum rod for electrical purpose *Journal of Rare Earth*, Vol.**24**, Spec. Issue, p.355.

[3] Akhmad Z 1997 Effect of Zirconium Addition on *ACSR Thesis S2, Metallurgy Engeneering, University of Indonesia*

[4] Zulkifli 1997 Research on The current Conductivity Ability for Heat Resistant Aluminum Conductors", *Thesis S2 Material Science, University of Indonesia*

[5] Roohollah J and Mohammad R.T 2010 Manufacturing of high-strength aluminum/alumina composite by accumulative roll bonding, *Materials Science and Engineering* A **527** pp 4146-4151.

[6] Takashi M and Hideaki M 2010 Effect Titanium Addition on Composite $L1_2$ Presipitates in Aluminum-Zirconium Alloy *Materials Science Forum* Vols.**638-642** pp 279-284

[7] ASTM standard: B 108/B 108M – 08, Standard Specification for Aluminum-Alloy Permanent Mold Castings

[8] A.A. Rao, B.S. Murty and M. Chakroborty, 1997 Role of zirkonium and impurities in grain refinement of aluminum with Al-Ti-B *Materials Science and Technology*, Vol.**13**

[9] Roohollah J and Mohammad R.T 2010 Manufacturing of high-strength aluminum/alumina composite by accumulative roll bonding, *Materials Science and Engineering* A **527** pp 4146-4151.

[10] Pradeep R 2006 *Metal Matrix Composite Nanoparticles (MMCs)*, University of Wisconsin-Milwaukee

Designing an appointment system for an outpatient department

ChalitaPanaviwat[1], Haruetai Lohasiriwat[1*], and Wipawee Tharmmaphornphilas[1]
[1]Department of Industrial Engineering, Faculty of Engineering, Chulalongkorn University, Bangkok, Thailand. 10330

Abstract. An outpatient department in which non-urgent patients arrived for consultation faces with the problem of patients' long waiting time. Lack of an effective appointment is considered as the cause. In this study, a discrete event simulation model was developed to evaluate an appointment system. The developed system considered appointment rule, sequencing rule and environmental factor (priority rule). Various appointment systems were examined by the simulation model. All proposed alternatives effectively reduced outpatient waiting time without adding extra resources. With the introduction of appointment system, simulation had shown that average waiting time could be reduced thirty-seven to forth-four percent as compared to the current status. The most effective appointment system was finally recommended to the outpatient department for implementation.

Keywords: Appointment System, Waiting Time, Outpatient Department, Simulation

1. Introduction

Outpatient department is a main healthcare service for non-urgent patients. Many hospitals have a common problem of long waiting time in this particular department. The patients' survey we conducted also showed that long waiting time is their most often encountered problem. It is also a major cause of patient dissatisfaction in outpatient department's service [1] particularly that the usual service time is only few minutes with doctor consultation. Through our observation and interviewing at a hospital in Thailand, we found a number of patients arrive in awfully early morning (i.e., 4.00am until 8:30am). They expect to get queue for doctor consultation at the earliest possible. Their normal understanding is that the earlier come, the faster finish. This attitude results in congestion at the outpatient service long before the service hour even starts. Unfortunately, with the attempt to overcome this problem, the hospital manages by introducing an earlier nurse schedule. Nurses at triage eventually have to work before their official working time. According to Soriano A [2], long waiting time in outpatient services is dispensable. Establishing appropriate appointment system is one solution to this problem. Basically, appointment system can better align patient arrival patterns to the doctors' and nurses' actual services rather than the vice versa.

An appointment system consists of appointment rules and sequencing rules. Appointment rules include appointment interval (the length of appointment slot), block size (the number of patients in each appointment slot) and initial block (the number of patients in the first block of appointment system). These three factors differentiate types of appointment rules such as a single block system, an individual block system and a multiple block system. The single block system appoints all patients by

* Corresponding Author: Haruetai Lohasiriwat, Email: haruetai.l@chula.ac.th

date, which all patients are assigned to the same time period. This appointment makes a long waiting time but many outpatient departments still use [3].The individual block/ fixed interval system gives a specific time with equal interval for each patient [4]. The individual block/ fixed interval with an initial block system is the same as the individual block but more than one patient in the first slot. For example, Bailey's rule suggests placing two patients in the beginning of the session [5].The multiple block/fixed interval is similar to the individual block. However, the multiple block assigns more than one patient to each slot. For sequencing rules, there are many policies such as FCFA (no sequencing rule), RTBG (service return patients first then follow by new patients) and ALTER (alternating pattern between new and return patients). Cayirli [6] found that RTBG and ALTER were the best sequencing rules tested. With various type of appointment system available, it should consider environmental factors (e.g., priority rules, punctuality of patient arrival, no-show rate and resource-sharing) before implementing an appointment system. A priority rule is an ordering queue for serving patients in each appointment block such as FCFS (first-come, first-served) and serving appointment patients firstly.

Many researches from the earlier studies considering an appointment system only at the appointment rules and finding includes number of appointment patients in each appointment slot and the length of appointment interval. Yeon [7] identified that an appointment system for multiple doctors is not the best appointment system for an individual doctor. A hospital in Japan assigning two patients to a block and doubling up consultation's service time per interval, reduction of patients waiting time is also found [8]. On the contrary, our study additionally analyzes sequencing rules and priority rules together with the appointment rules to examine effects on the overall patients' waiting time. Our main objective is to develop a detailed appointment system for the case study hospital that could reduce patient waiting time without adding extra resource(s).

2. Model

2.1. Description of an outpatient department

The outpatient department in this study has four clinics which are surgery, general practice, medicine and obstetric. The four clinics share common resources including nurses and laboratory test. This outpatient department operates with two types of patients: appointment (A) and non-appointment patients (NA). Non-appointment patients include new patients who first arrive at the hospital, walk-in patients and no-show patients who missed their appointments.

Patient flow of the outpatient department is shown in figure1. All appointment patients must go to a reception deck to identify the appointed clinic and check if there is any laboratory test required. Patients who need laboratory tests can go directly for the test and later come back for triage before doctor consultation. Otherwise, patients go for triage and follow by consultation. All non-appointment patients, on the other hand, need to pass registration process. Next step is screening by nurses to assign for the doctor clinic. In case of laboratory tests required, patients go for testing before consultation similar to the appointed group. After triage, all patients are queued for consultation. Then, according to doctor's recommendation, some patients could need additional laboratory tests or may be referred to other specialists. These patient groups usually assigned to come back to the same consulted doctor. After all consultation processes completed, patients may make an appointment for the next visit and get drugs from a pharmacy before leaving the hospital. These last two services are out of our scope.

2.2. Performance Measure

The objective of this study is to reduce patients' waiting time in the outpatient department by developing an appointment system that matching with doctors' and nurses' services available.

IOP Conf. Series: Materials Science and Engineering **58** (2014) 012010 doi:10.1088/1757-899X/58/1/012010

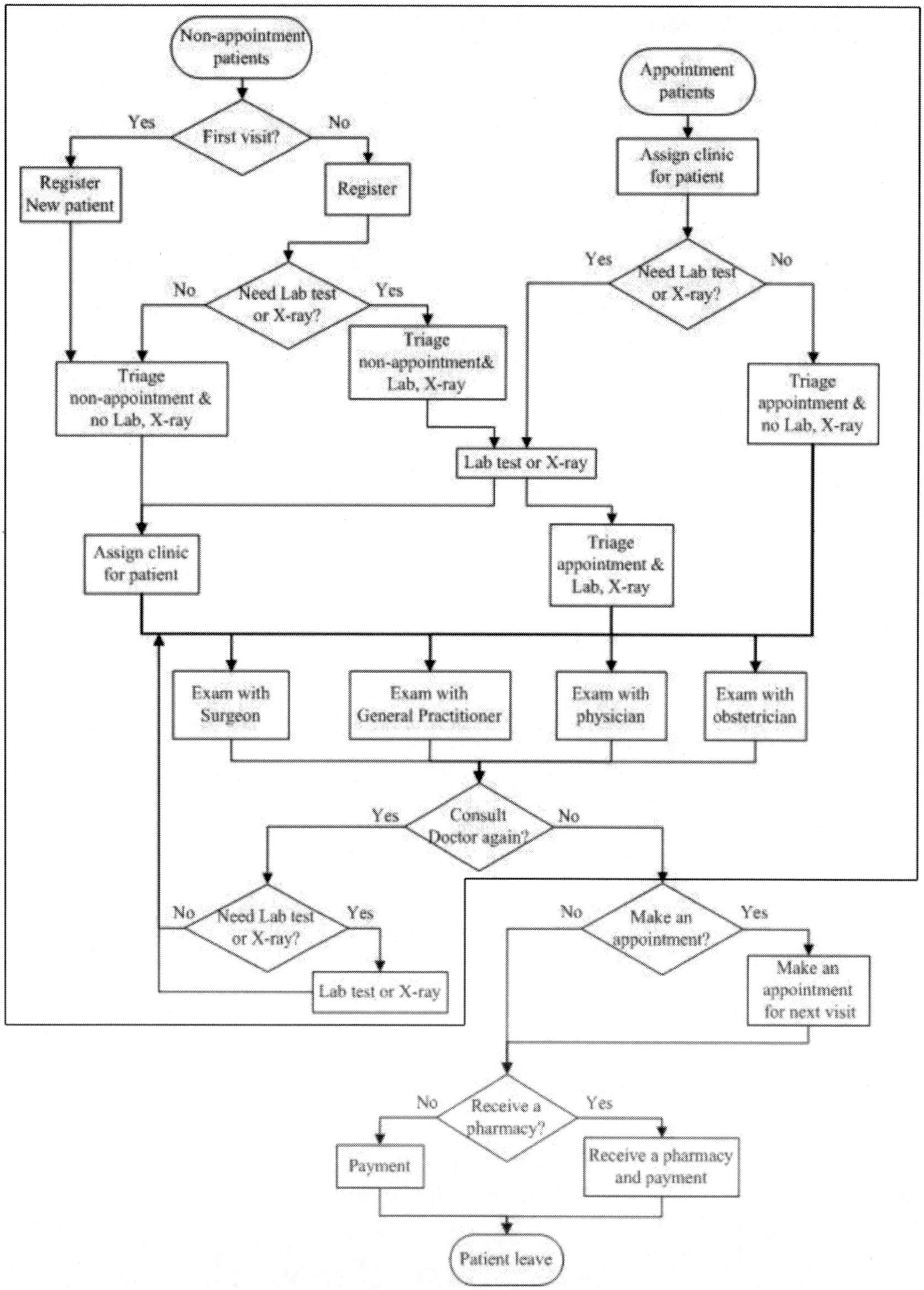

Figure 1. Patient flow diagram (scope of the study is shown within the designated line)

The performance measure to evaluate appointment system is a weighted average waiting time (WAWT). This measure is calculated using equation (1).

$$\text{WAWT} = \frac{(\text{AWTA}\cdot\text{NA})+(\text{AWTNA}\cdot\text{NNA})}{\text{NA}+\text{NNA}} \tag{1}$$

Where AWTA=Average waiting time for appointment patients, NA=Number of appointment patients, AWTNA=Average waiting time for non-appointment patients, NNA=Number of non-appointment patients

2.3. Data Collection and Simulation Model

The outpatient department opens Monday to Friday from 7:00 am until last patient leaves the department. However, the consultation hours are 9:00-12:00 am only. The research team manually tracks the actual time patient arrived and left each service station for a week. Patient arrival patterns from Monday to Thursday are similar, but Friday is different since surgery clinic is closed. Therefore, data used in this version of a simulation model is the one collected from Monday to Thursday period.

We studied 4 clinics (surgery, general practice, medicine, and obstetric) which are in the same building area. There is a doctor per clinic from Monday to Thursday. Four nurses work mainly in triage and two staffs work in registration. These are common resources for all clinics. All human resources involved in the outpatient department are shown in table 1.

Table 1. Resource of doctors and nurses.

Resource	Number
Registration	2 staffs
Triage for appointment	2 nurses
Triage for non-appointment	2 nurses
Surgeon	1 doctor
General practitioner	1 doctor
Physician	1 doctor
Obstetrician	1 doctor

Inter arrival time for appointment and non-appointment patients are shown in table 2.Service times were additionally collected between May 27-28th and June 5-7th, 2013. Approximately 400 patients were observed. The summary of service times are shown in table 3.

Table 2. Inter arrival time for patients.

Patient Category	Arrival time for patients	Inter Arrival Time (seconds)
Appointment patients	07:00-8:00 am	47 + EXPO(139)
	08:00-9:00 am	50 + EXPO(267)
	09:00-10:00 am	30+EXPO(508)
	10:00-12:00 am	315 + EXPO(280)
Non-appointment patients	07:00-8:00 am	179 + EXPO(253)
	08:00-9:00 am	51 + EXPO(161)
	09:00-10:00 am	2 + EXPO(253)
	10:00-12:00 am	177 + EXPO(328)

A simulation model for the outpatient department was developed using the ARENA Software version 14.0. In our observation, triages and consultation room were the most congested units. Therefore, the model is focused only from patient arrival until consultation is completed and included all pertinent resources.

Table 3. Service time in each unit.

Service time	Time (seconds)
Registration for new patient	300 (constant)
Registration	180 (constant)
Triage for appointment and lab test patient	100 + EXPO(109)
Triage for appointment patient	100 + WEIB(88.2, 1.05)
Triage for non-appointment and lab test patient	110 + WEIB(134, 0.569)
Triage for non-appointment patient	151 + GAMM(66.6, 1.23)
Surgeon's consultation time	85 + WEIB(136, 0.882)
General practitioner's consultation time	78 + WEIB(179, 1.11)
Physician's consultation time	120 + EXPO(120)
Obstetrician's consultation time	123 + WEIB(371, 0.601)
Lab test- 1st time [a]	890 + 9.07e+3BETA(0.853, 1.27)
Lab test- 2nd time	795+4.65e+3BETA(0.74,1.25)

[a] Waiting time is included in the service time of this unit
EXPO=Exponential, GAMM=Gamma, WEIB=Weibull

2.4. Verification and validation of the model
Model verification and validation were performed. Figure 3 shows waiting time comparison between actual data and results from running 30 replicates of the model. Results from t-tests show that waiting time for each unit in model and actual data are not significantly different at the 95% confidence interval.

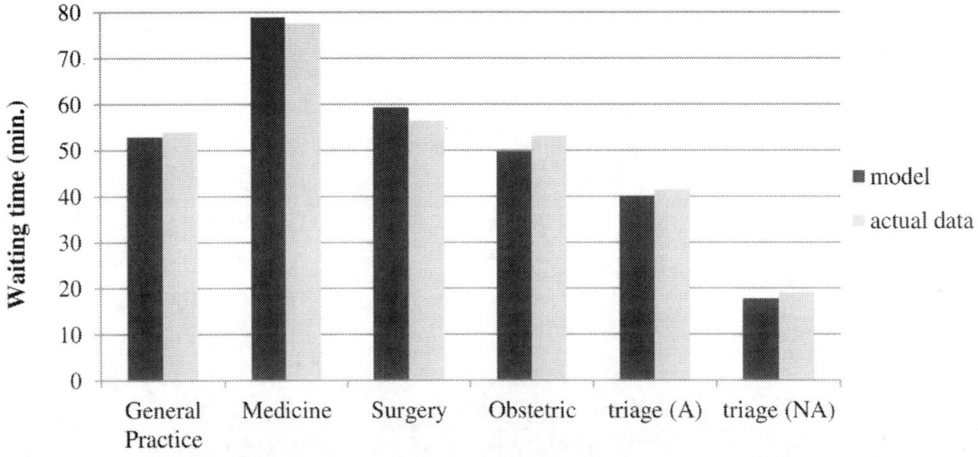

Figure 3. Waiting time for each unit from model and actual data.

Figure 4 shows comparison of weighted average waiting times. In this study, average waiting time for appointment patients is about 40% more than non-appointment patients.Results from t-tests show that AWTA, AWTNA and WAWT (weighted average of waiting time) calculated from the model and the actual data are not significantly different at the 95% confidence interval. From the statistical point, the simulation model hence can be used to simulate the real setting.

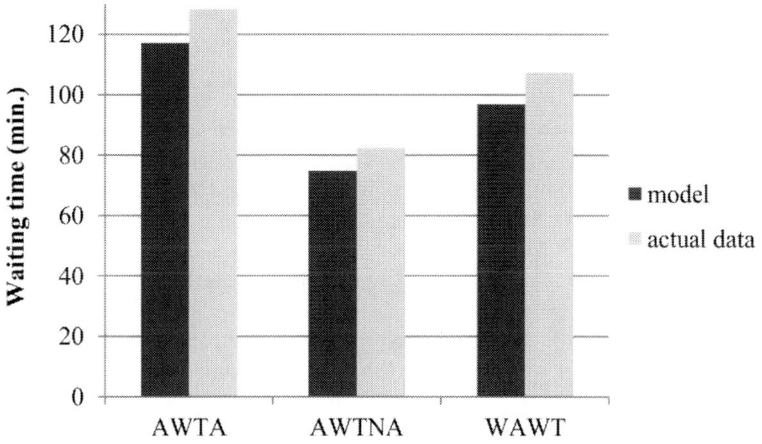

Figure 4. Weighted average waiting time.

3. Experiment

We study appointment systems by evaluating appointment interval, block size, sequencing rules, and priority rules. We experiment 30 and 60 minutes interval which come from surveying patients' preferences via direct interview. The block size for each clinic is calculated using an average consultation time. In an appointment, appointment patients can classify to patients who have a laboratory test and no laboratory test. A sequencing rule in this study is to order appointment patients who have a laboratory test in each clinic first then follow by appointment patients who have no laboratory test. Two priorities, first-come first-served (FCFS) and serving appointment patient first then non-appointment patient, are tested. In conclusion, there are four alternative scenarios in the experiment as listed below.

Alternative 1: 30 minutes interval. The numbers of patients in each block for surgery, medicine, general practice, medicine, and obstetric are 4, 4, 2, and 1 respectively. FCFS priority rule is tested.

Alternative 2: 30 minutes interval. The numbers of patients in each block for surgery, medicine, general practice, medicine, and obstetric are 4, 4, 2, and 1 respectively. Priority, serving appointment patients firstly, is tested.

Alternative 3: 60 minutes interval. The numbers of patients in each block for surgery, medicine, general practice, medicine, and obstetric are7, 7, 4, and 1 respectively. FCFS priority rule is tested.

Alternative 4: 60 minutes interval. The numbers of patients in each block for surgery, medicine, general practice, medicine, and obstetric are 7, 7, 4, and 1 respectively. Priority, serving appointment patients firstly, is tested.

After running 30 replications for each alternative, the results are shown in table 4. Using 30-minute interval, average waiting time is about 54 minutes which is approximately 44% reduced from the base case (existing system). Using 60-minute interval, average waiting time is about 60 minutes which is approximately 37% reduced from the base case.

Table 4. Weighted average waiting time for each schedule.

Scenarios	WAWT (min.)	% of WAWT Reducing
Base	96.94	0
Alternative 1	54.05	44.25
Alternative 2	53.97	44.33
Alternative 3	60.45	37.64
Alternative 4	60.17	37.93

Examining in more detail, we found some different between appointment and non-appointment patients as shown in table 5. FIFO priority in alternatives 1 and 3 lead to the waiting time reduction of both appointment and non-appointment patients. While priority, serving appointment patients firstly, in alternatives 2 and 4 reduce waiting time for appointment patients more than non-appointment ones. Note that, in all four alternatives we tested, resources' utilization such as staffs, nurses and doctors are close to the existing system. As a result, resources in any appointment system are still effective.

Table 5. Waiting time and percent of reduction for each alternative.

Scenarios	AWTA(min.)	%AWTA Reduced	AWTNA(min.)	%AWTNA Reduced
Base	117.10	0	74.79	0
Alternative 1	51.77	55.79	56.41	24.58
Alternative 2	37.46	68.01	70.44	5.82
Alternative 3	59.46	49.22	61.50	17.77
Alternative 4	46.34	60.42	74.80	-0.02

4. Conclusion

We study an outpatient department in a Thai hospital, which faces patients' waiting time problem. With the current single block system, an average waiting time of appointment patients are 40% longer than that of non-appointment patients. Consultation rooms and triages are the most congested areas. Our proposed appointment system can effectively manage patients matching with nurses and doctors' service. All multiple block, fixed interval appointment systems proposed in this paper outperform the current system. The results show that 30-minute appointment intervals reduce approximately 44% of average waiting time while 60-minute appointment intervals reduce approximately 37%. Priority, serving appointment patients firstly, focuses on reducing average waiting time for appointment patients comparing to FCFS priority where types of patients are not affected. Our final proposed schedules are shown in table 6 and 7. Note that since the 30-minute and 60-minute interval seems to show only slight different (approximately 10 minutes), we support the use of either appointment schedule.

Table 6. Appointment schedule for 30-minute interval.

Consultation's appointment time	Number of patients each clinic			
	Surgery	General Practice	Medicine	Obstetric
9:00 – 9:30	4[a]	2[a]	4[a]	-
9:30 – 10:00	4	2[a]	4[a]	-
10:00 – 10:30	4	2	4	1[a]
10:30 – 11:00	4	2	4	1
11:00 – 11:30	4	2	4	-

[a]Four patients with laboratory test.

Table 7. Appointment schedule for 60-minute interval.

Consultation's Appointment time	Number of patients each clinic			
	Surgery	General Practice	Medicine	Obstetric
9:00 – 10:00	7(4,3)[a]	4[b]	7[b]	-
10:00 – 11:00	7	4	7	1[b]
11:00 – 12:00	7	4	7	1

[a] Four patients with laboratory test and three patients without laboratory test.
[b] Four patients have a laboratory test.

For implementation, there will be some adding steps to make appointment according to our proposed model(s). First, the few earliest slots are exclusive for appointed patients. Then, for priority (ordering patients' queue after triage), we support the FCFS (first come first serve) due to the fairness between appointed and non-appointed groups. However, we leave final decision on the hospital management team to take into account other consideration such as whether there should be preference for appointed patient. In any circumstances, there will be extra requirement to track each patient's arrival time to the department. Also, among appointment patients, ones who have a laboratory test should have first priority for doctor's consultation. Then, the patients who arrive on appointment time have second priority. Last priority will then go to unpunctual patients and non-appointment patients. Normal practice of requesting patient to arrive before appointed time is enforced (i.e., at least one hour for patient with laboratory test and half hour for non-laboratory one).

Acknowledgement
The authors wish to express sincere gratitude to Thai Health Promotion Foundation for supporting our research. The authors would like to thank Thira Woratanarat, M.D. for his valuable comments, and also thank staffs in Damnoensaduak hospital for data collection, site visit and other various supports.

References

[1] Huang X 1994 Patient attitude towards waiting in an outpatient clinic and its application *Health Services Management Research* **7** 2-8
[2] Soriano A 1966 Comparison of two scheduling systems *Operation Research* **14** 388-397
[3] Cayirli T and E Veral 2003 Outpatient scheduling in health care: a review of literature *Production and Operations Management* **12** 519-549
[4] Klassen K J and T R Rohleder1996 Scheduling outpatient appointments in a dynamic environment. *Journal of Operations Management 14* **2** 83-101
[5] Bailey N 1952 A study of queues and appointment systems in hospital outpatient departments with special reference to waiting times *Journal of the Royal Statistical Society* 14 **2** 185-199
[6] Cayirli T, E Veral and Rosen H 2006 Designing appointment scheduling systems for ambulatory care services*Health Care Management Science* **9** 47-58
[7] Yeon N, Lee T and Jang H 2010 Outpatients appointment scheduling with multi-doctor sharing resources *Proceedings of the Winter Simulation Conference* 3318-3329
[8] Wijewickrama A K A and S Takakuwa 2005 Simulation analysis of appointment scheduling in an outpatient department of internal medicine *Proceedings of the Winter Simulation Conference* 2264-2273

The impacts of final demand changes on total output of Indonesian ICT sectors: An analysis using input-output approach

Ubaidillah Zuhdi

Department of Industrial Administration, Tokyo University of Science, Noda 278-8510, Japan
School of Business and Management, Institut Teknologi Bandung, Bandung 40132, Indonesia

ubaidillah.zuhdi@gmail.com

Abstract. The purpose of this study is to analyze the impacts of final demand changes on total output of Indonesian Information and Communication Technology (ICT) sectors. This study employs Input-Output (IO) analysis as a tool of analysis. More specifically, demand-pull IO quantity model is applied in order to achieve the objective. "Whole sector change" and "pure change" conditions are considered in this study. The results of calculation show that, in both conditions, the biggest positive impact to the total output of the sectors is given by the change of households consumption while the change of import has a negative impact. One of the recommendations suggested from this study is to construct import restriction policy for ICT products.

1. Introduction

Technology is undoubtedly a tool used almost every day by people. For example, we can see that people use transportation devices, such as car, motorcycle and train, when they are going to somewhere. Moreover, people also use technology when they do their hobbies, such as photography, traveling, and game. Such examples affirm that, in recent days, technology could not be separated from daily life of society.

The importance of technology is also mentioned by researchers. [1] argued that now technology impacts, and impacted by, nearly every aspect of human life. [2] explained that technology is one of the major factors that dictate the history path and society form. [3] mentioned that green technology will have an important role in generating the power as countries search to reduce the dependence on the fuels which are imported, reinforce their independence on energy, and decrease the use of fuels that cause pollution.

One of the technologies frequently used by people is an Information and Communication Technology (ICT). The use of ICT can be easily observed in recent years. For instance, we can easily see people use mobile phone as well as fixed one when they communicate with others. The similar phenomenon can be seen on the use of internet. Nearly every aspect of society now needs internet. In other words, internet has become a powerful tool both in micro and macro activities. [4] affirmed this phenomenon, namely millions of people consume their hours each day for accessing and making information online.

There are many previous studies discussed about ICT. [5] explored the effect of ICT investments on development of human. More specifically, they investigated the correlations between different dimensions of ICT investment and the aspects of human development. Their objects of study consisted of 51 countries which were divided to three groups, namely high, middle, and low income countries. [6] developed the definitions and applications hierarchies for the term of ICT. [7] assessed the role of ICTs in enhancing access to agricultural information which was needed to develop the agribusiness and agricultural production. Their study focused on Tanzania.

[8] examined the correlation of investments in ICT and growth of economy in the geographical configuration using Generalize Method of Moments (GMM). Their study focused on OPEC member

countries. The period of analysis of their study was from 1990-2007. [9] investigated the role of ICT sectors in Indonesian national economic structural changes using Structural Decomposition Analysis (SDA), one of the analysis tools in Input-Output (IO) analysis. Their analysis period was from 1990-2005. [10] explored how to encourage ICT sectors of Indonesia using IO approach.

On the other hand, [11] compared the role of ICT sectors in the national economic structural changes of Japan and Indonesia using SDA. [12] employed two methods, SDA and simple output multipliers, in analyzing the role of ICT sectors in Japanese national economy from 1995-2005. He compared the results of calculation using these methods in order to get another perspective related to this role. The study discusses the impacts of changes of final demand, one of the variables in economy, on total output of ICT sectors of specific country, however, is still thin. The kind of study is important because it can describe more detail the characteristics of these sectors. Further, this description can be used to inquire the ways in improving the sectors. This study is conducted to fulfill this gap.

The purpose of this study is to analyze the impacts of changes of final demand on total output of ICT sectors of particular country. Analysis in this study focuses on Indonesia. In order to achieve the objective, this study employs IO analysis. This paper is arranged as follows. Chapter 2 explains the literature related to method which is used in this study. The methodology of this study is discussed in chapter 3. Chapter 4 explores the results of calculation and analysis of these results. Chapter 5 describes the conclusions of and further researches suggested from this study.

2. Literature review

The following description is fundamental information in IO analysis which is described by [13]:

"*The fundamental information used in IO analysis concerns the flows of products from each industrial sector, considered as a producer, to each of the sectors, itself and others, considered as consumers. The basic information from which an IO model is developed is contained in an interindustry transactions table.*"

[14] explained that IO analysis is an economic tool that determines the connections between industrial sectors in the economy. They also mentioned that IO tables describe the inputs of commodity that are employed by each industrial sector to generate its output, the commodities made by each industrial sector, and the utilization of commodities by final consumers.

Demand-pull IO quantity model is one of the analysis tools in IO analysis. [13] argued that, in this model, the change appears on the quantity of commodities while the prices are fixed. Final demand acts as an exogenous variable while this model is applied.

3. Methodology

As described in introduction, the purpose of this study is to analyze the impacts of final demand changes on total output of Indonesian ICT sectors. IO analysis is employed in this study. More specifically, demand-pull IO quantity model is applied to achieve the objective. Following explanation describes the methodology of this study.

The first step is to define the data of this study. Adjusted and aggregated IO table of Indonesia for 2005 is used as data. The second step is to describe the ICT sectors used in this study. Table 1 shows these sectors. The next step is to conduct the calculation in order to know the impacts of modifications of final demand on total output of Indonesian ICT sectors. Demand-pull IO quantity model is applied in this calculation. According to [13], following equation describes this model:

$$\mathbf{x}^1 = \mathbf{L}^0\mathbf{f}^1 \tag{1}$$

where x, L, and f are matrices of total output of sectors, Leontief inverse, and final demand of sectors, respectively. *0* and *1* indicate initial and future times, respectively. Initial period in this calculation is 2005. The scenarios of final demand modification used in this study are described in table 2.

Table 1. ICT sectors of Indonesia used in this study.

No.	Sector Number	Sector Name
1	135	Construction and installation on electricity, gas, water supply, and communication
2	146	Communication services

(Source: [11])

Table 2. The scenarios of final demand modification used in this study.

Component of final demand	Scenario 1	Scenario 2	Scenario 3
	Export modification	Import modification	Households consumption modification
Export	Increases 30%	Constant	Constant
Import	Constant	Increases 30%	Constant
Households consumption	Constant	Constant	Increases 30%

(Source: [15] with slight modifications)

"Whole sector change" and "pure change" conditions are considered in the calculation. Former word explains the condition which modifications of final demand are addressed to all Indonesian industrial sectors while latter one only focuses on ICT sectors of Indonesia. In this study, former one will be called condition A while condition B is used in describing latter condition. Recommendations are suggested based on the calculation results. Conclusions of this study and suggestions related to the further research are described on final step.

4. Results and analysis

Figures 1 and 2 describe the total output of Indonesian ICT sectors for each scenario on condition A. X_t denotes total output of these sectors on initial period while total output on future period are explained by X_{t+1}. Based on information in these figures, in this condition, scenario 3, the change of households consumption, has the biggest positive impact to the total output of both Indonesian ICT sectors. On the other hand, scenario 2, the modification of import, gives negative impact to the total output of these sectors.

Figures 3 and 4 show the total output of Indonesian ICT sectors for each scenario on condition B. As with the previous explanation, X_t explains the total output of these sectors on initial period while X_{t+1} describes the total output on future period. These figures show that, in this condition, scenario 3, households consumption modification, has the biggest positive impact to the total output of both Indonesian ICT sectors. Contrarily, scenario 2, the change of import, generates the negative impact to the total output of these sectors.

Based on previous explanations, several similarities appear on both conditions. These are 1) the biggest positive impact to the total output of Indonesian ICT sectors is delivered by scenario 3, the change of households consumption, and 2) scenario 2, the change of import, gives the negative impact. These characteristics will be used as a foothold in suggesting the recommendations for above sectors.

[10] described the suggestions in improving ICT sectors of Indonesia, namely 1) to implement broadband internet service especially on the dense area, 2) to improve the mobile telecommunication access quality, 3) to improve the national postal service, 4) to improve the broadcasting services, and 5) to improve the activities related to the ICT commodities export. These actions focus on the improvements of ICT sectors so the demand from households and activities of export of these sectors is expected to increase.

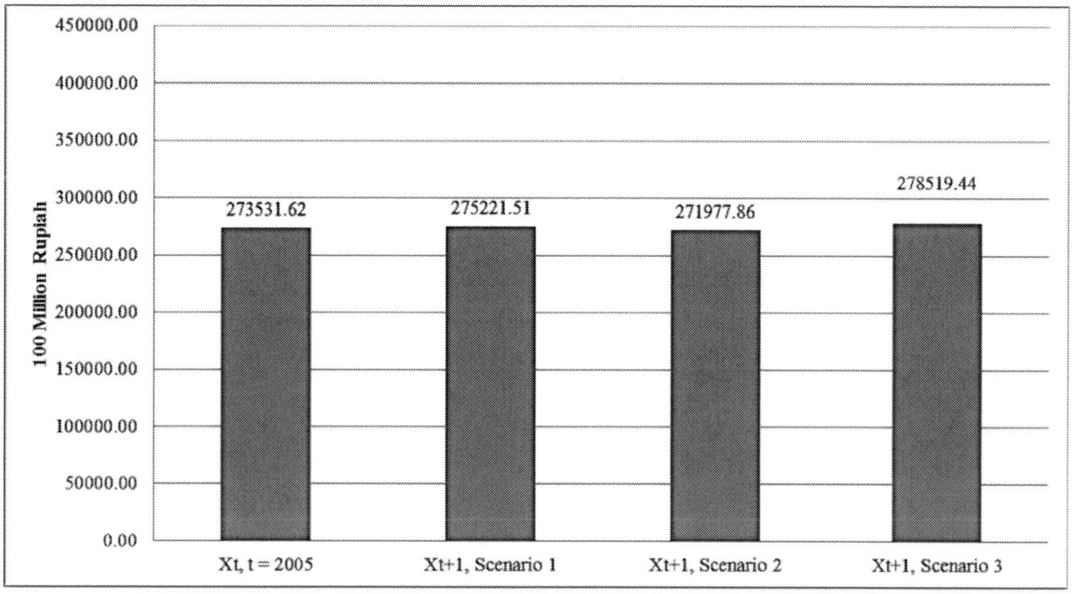

Figure 1. Total output of construction and installation on electricity, gas, water supply, and communication sector for each scenario on condition A

(Source: [10] with slight modifications).

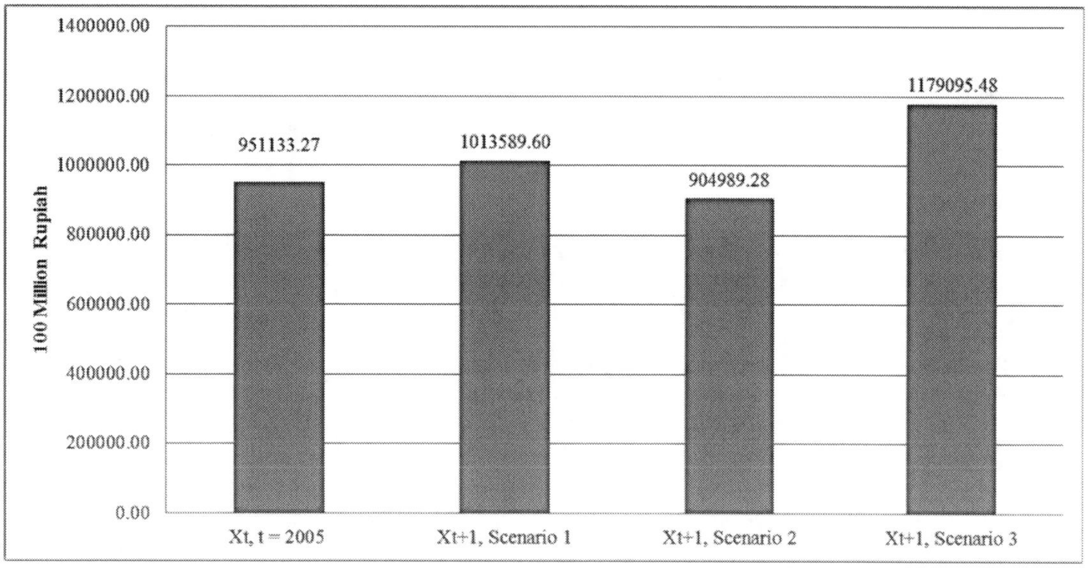

Figure 2. Total output of communication services sector for each scenario on condition A

(Source: [10] with slight modifications).

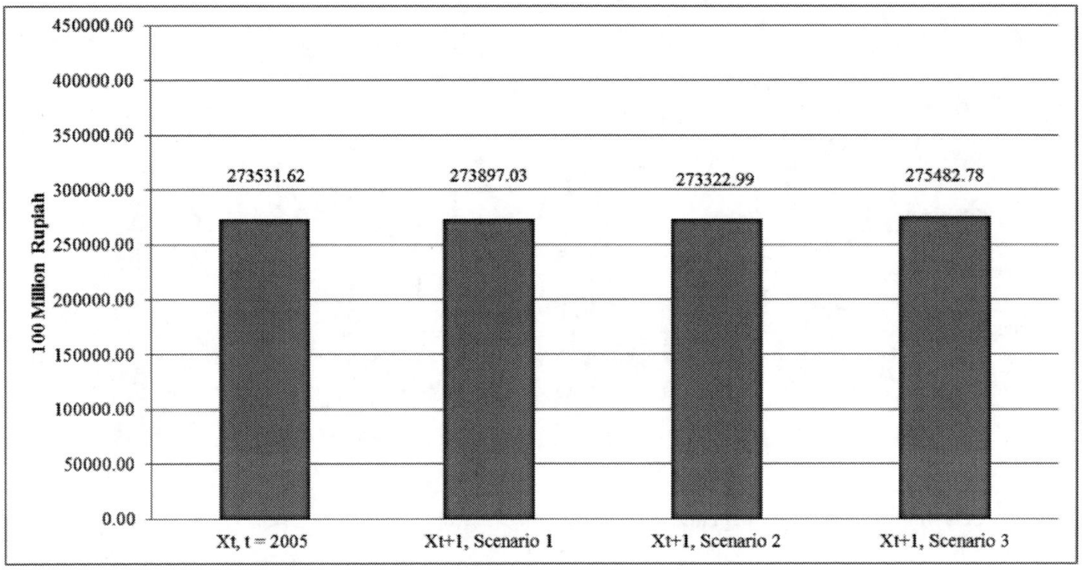

Figure 3. Total output of construction and installation on electricity, gas, water supply, and communication sector for each scenario on condition B.

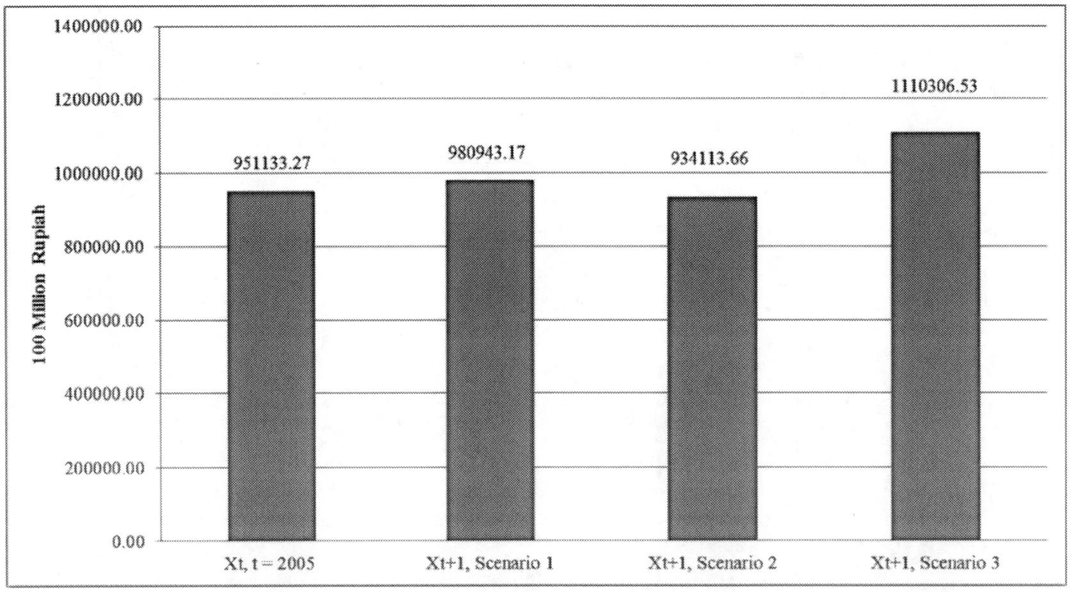

Figure 4. Total output of communication services sector for each scenario on condition B.

The maneuver related to the activities of import of ICT sectors, however, has not been discussed. This maneuver is needed because, based on the previous explanations, the change of import will give negative impact. In other words, import activities related to the products of ICT should be avoided if the increasing of total output of above sectors is expected in future period.

Above logic is strengthened by following explanation. Import activities tend to delimitate industrial sectors in producing more outputs. This situation will be worse if the import products have higher competitiveness in the market. The consequence of this circumstance is the decreasing of total output of the sectors in future period. Therefore, the policy related to the import restriction on ICT products is needed in order to make sure the enhancement of total output of ICT sectors.

Obviously, the import activities are still needed by ICT sectors of Indonesia. This fact seems especially on the ICT products which these sectors do not produce. In other words, the import restriction policy should focus on the products of ICT which above sectors have an ability to produce.

5. Conclusions and further research

This study analyzed the impacts of final demand changes on total output of Indonesian ICT sectors using demand-pull IO quantity model. The results of calculation showed that, in analyzed conditions, these sectors had similarities. These were the biggest positive impact to the total output of the sectors was given by the change of households consumption while the change of import had a negative impact. Recommendations in improving Indonesian ICT sectors were compiled based on these characteristics. These were 1) to implement broadband internet service especially on the dense area, 2) to improve the mobile telecommunication access quality, 3) to improve the national postal service, 4) to improve the broadcasting services, 5) to improve the activities related to the ICT commodities export, and 6) to construct import restriction policy for ICT products. This policy should focus on ICT products which above sectors have the capability to generate.

The discussion about the price change in ICT sectors, however, is not explored in this study. In IO approach, analysis of this change will show the impacts of modifications of value added parts on the price of these sectors. In other words, this analysis is also important because it can describe more detail the dynamics of the sectors. Therefore, this analysis is suggested as a further research. Besides, international comparison in the topic of this study, especially comparison among developing and developed countries, will also be an attracting further research. This comparison will show the characteristics of ICT sectors of compared countries.

References

[1] Grübler A 1998 *Technology and Global Change* (Cambridge: University Press)
[2] Braun E 2010 The changing role of technology in society *http://epub.oeaw.ac.at/ita/ita-manuscript/ita_10_03.pdf* Accessed 19 October 2013
[3] Abraham D S 2012 The battle for new resources: Minor minerals in green technologies *http://www.rieti.go.jp/jp/publications/pdp/12p005.pdf* Accessed 19 October 2013
[4] Peitz M and Waldfogel J 2012 *The Oxford Handbook of the Digital Economy* (Oxford: University Press)
[5] Bankole F O, Shirazi F and Brown I 2011 Investigating the impact of ICT investments on human development *The Electronic Journal of Information Systems in Developing Countries* **48** 0119
[6] Zuppo C M 2011 Defining ICT in a boundaryless world: The development of a working hierarchy *International Journal of Managing Information Technology* **4** 1322
[7] Mtega W P and Msungu A C 2013 Using information and communication technologies for enhancing the accessibility of agricultural information for improved agricultural production in Tanzania *The Electronic Journal of Information Systems in Developing Countries* **56** 0114
[8] Nasab E H and Aghaei M 2009 The effect of ICT on economic growth: Further evidence *International Bulletin of Business Administration* **5** 4656
[9] Zuhdi U, Utomo D S and Alamanda D T 2011 Analyzing the role of ICT sector to the national economic structural change: The case of Indonesia *Jurnal Manajemen Teknologi* **10** 299307
[10] Zuhdi U, Putranto N A R and Prasetyo A D 2014 Encouraging information and communication technology sectors using input-output approach: The case of Indonesia *Advanced Science Letters* **20** 199202

IOP Conf. Series: Materials Science and Engineering **58** (2014) 012011 doi:10.1088/1757-899X/58/1/012011

[11] Zuhdi U, Mori S and Kamegai K 2012 Analyzing the role of ICT sector to the national economic structural changes by decomposition analysis: The case of Indonesia and Japan *Procedia-Social and Behavioral Sciences* **65** 749754

[12] Zuhdi U 2014 The other perspective related to the role of information and communication technologies sectors in national economy: The case of Japan *Advanced Science Letters* **20** 483486

[13] Miller R E and Blair P D 2009 *Input-Output Analysis: Foundations and Extensions* (Cambridge: University Press)

[14] Horowitz K J and Planting M A 2006 Concepts and methods of the input-output accounts *http://www.bea.gov/papers/pdf/IOmanual_092906.pdf* Accessed 9 March 2014

[15] Zuhdi U, Prasetyo A D and Sianipar C P M 2013 Analyzing the dynamics of total output of Japanese creative industry sectors: An input-output approach *Procedia Economics and Finance* **5** 827835

Improvement of Project Portfolio Management in an Information Technology Consulting Company

S Kaewta[1] and **P Chutima**[2]
Faculty of Engineering/Industrial Engineering
Chulalongkorn University, Wangmai, Patumwan, Bangkok 10330, Thailand

E-Mail: [1]Sirirat_kaewta@hotmail.com; [2]cparames@chula.ac.th

Abstract. The scope of this research is to improve the efficiency of multiple project management in an information technology consulting company through the adaptation of the project portfolio management technique. The project management information system (PMIS) is implemented to establish effective communication channels so that internal and external teams as well as all relevant stakeholders can be employed to negotiate their work schedules. In addition, all activities created by multiple teams can be systematically reviewed and combined into a single checklist to be used as an agreed working plan for all team members. A general guideline for project portfolio management in information technology consulting companies is also proposed, which could results in a higher level of project on-time delivery.

Keywords: Project Portfolio Management, Communication Plan, Standard Procedures

1. Introduction

The case study of this research is an information technology consulting company which manages and customizes IT solutions according to the needs from multiple industries such as government office, banking, corporate company, university, etc. The head quarter of the case study is located in Singapore, managing multiple projects in South East Asia including Thailand, Malaysia, Philippines, Indonesia as well as such emerging markets like Sri Lanka, Bangladesh, Brunei and Vietnam.

One of the key performance indicators of the case study is to successfully manage multiple projects to achieve project on-time deliveries and hence, the percentage of project on-time delivery is its performance measurement. The annual target for every country is to achieve at least 95% of the projects being completed on-time. Statistically, during 2012 – 2013, 592 projects were launched in South East Asia. However, the number of projects completed on-time was only 532, resulting in 60 uncompleted projects which were carried over to 2014. Out of this number, 15 projects belong to Thailand.

Regarding the ongoing projects in Thailand, 8 uncompleted projects were launched in 2012 resulting in the percentage of on-time delivery in 2012 being much less than target. In addition, more new projects were added in 2013. Due to the low percentage of project on-time delivery, two main problems are identified by top management. The first problem is hardware related problem cause by the existing IT systems of customers that cannot properly interface with the new deployed technology. This problem has already been alleviated by providing a number of technical trainings and hardware solution supports. The second problem, is internal management related problem cause by poor IT specialist resource arrangement, miscommunication between sellers and project teams, together with vague standard operating procedures. The latter problem is the issue considered and discussed in this research.

The objective of this research is to improve the case study company's performance in managing multiple projects through adapting project portfolio management techniques based on the project management information system (PMIS) theory in order to establish effective communication plan and standard operating procedures that will be followed by all concerned parties.

Project management institute, as cited in "A Guide to the Project Management Body of Knowledge" (2011), defines project portfolio management as the centralized management of one or more portfolios, which includes identifying, prioritizing, authorizing, managing, and controlling projects, programs and other related works, to achieve specific strategic business objectives. Portfolio management ensures that projects and programs are reviewed to prioritize resource allocation. Also, the management of the portfolio is consistent with and aligned to organizational strategies. Project portfolio management adapts information technology as a tool to gather, integrate and disseminate the outputs of project management processes. It is used to support all aspects of the project, from initiation to completion, and it is applicable to both manual and automated systems.

Satawin (2011) adapted the techniques of project portfolio management to improve the production outputs of a construction company in order to meet delivery deadlines. He also developed the framework for resource allocation for current projects and resource planning for future projects. Malek and Alain (2011) used the Genetic Algorithm for fuzzy resource leveling for the helicopter maintenance domain. Marjolein and Ralph (2011) suggested project managers to apply PMIS as a decision support tool for planning, organizing, and controlling projects. Lee and Yu (2012) used PMIS to insure project success in construction. They developed and validated the application services providers and provided users with a useful framework for evaluating the success of PMIS. Braglia and Frosolini (2012) proposed PMIS to enhance planning and scheduling capabilities within a supply chain event management system to be able to manage the shipbuilding industry.

Project communications management includes processes required to ensure timely and appropriate generation, collection, distribution, storage, retrieval, and ultimate disposition of project information (PMBOK 2011). Communication management covers cross-cultural communication management during multicultural construction projects. Case studies from Kenya and UK indicated that one of the critical components of communication management was the creation and development of effective cross cultural collectivism, trust, communication and empathy in leadership, as well as execution of outcome (Ochieng and Price 2008). Rodney (2004) recommended that the best project performance occurred when there exists a high collaboration between the client and project manager, which could be realizable by incorporating the followings: (1) communication, (2) maintaining face-to-face meetings and (3) providing quantitative data required by the client.

2. Research Methodology

2.1 Project Management Information System

To increase the percentage of project completion and resource leveling, PMIS consisting of tools and techniques which are used to gather, integrate, and disseminate the output of project management process is applied. It is used to support all aspects of the project from initiation to closure (PMBOK 2011). Multiple projects details are gathered, such as project name, activity, period, resource, etc. Microsoft Project 2013 is used as a tool to manage multiple projects by inserting subprojects into a master project and pulling available resources from resource pool file (Bonnie 2013). The output of this method is a performance report which will provide various performance results including, but not limited to, the percentage of portfolio project completion, deliverable status, schedule progress, and resource usage.

2.2 Portfolio Communication Management

Portfolio communication management plan is developed by preparing inputs which are portfolio project details and portfolio reports such as percentage of portfolio project completion, resource utilization, issue items or risks, and other performance reports. Stakeholder analysis is used to identify recipients and set different groups of people separated by their concerns, interests, and actions. In addition, communication requirement analysis is applied to help determine proper frequency and communication vehicles. Finally, a portfolio management plan is developed to specify communication areas, frequency, intended recipients, and communication vehicles. This plan is approved by an authorized person.

2.3 Standard Guideline

The standard guideline is created through checklists related to the project management process, such as initial project, sales to delivery, internal meeting, performance monitoring, and closure. This document can help reduce duplicated jobs, rework and possibly reduce change requested from customers, and help the whole team to follow the standard operating process.

3. Result

Through the application of project portfolio management techniques, a standardized procedure which consists of performance reports, resource balance, new communication schedules, and checklists is created. Hence, the procedure increases the percentage of project delivery as shown below.

3.1 Result from Developed Portfolio Project Management

After applying Microsoft Project 2013 as a PMIS tool, project information can be presented to the team using graphical reports easy to customize. Fig. 1 shows the percentage of master project completion and the percentage of each non-master project completion within the same tracking period (week 13 was used for this data collection.) Information on the percentage of completion completed each week can then be collected and presented in a graphical report as shown in Fig. 2.

IOP Conf. Series: Materials Science and Engineering **58** (2014) 012012 doi:10.1088/1757-899X/58/1/012012

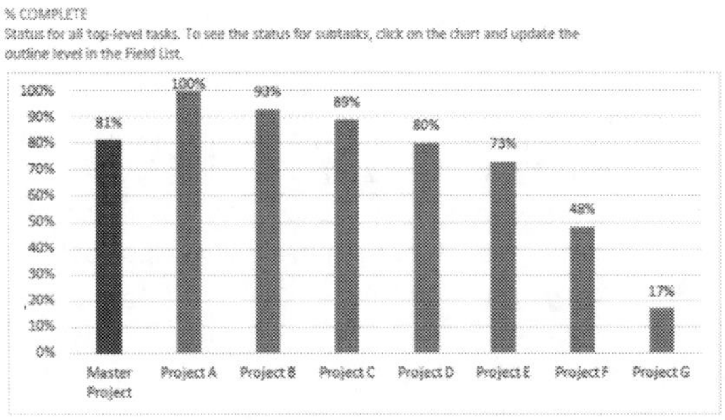

Figure 1. Percent Projects Complete at week 13rd

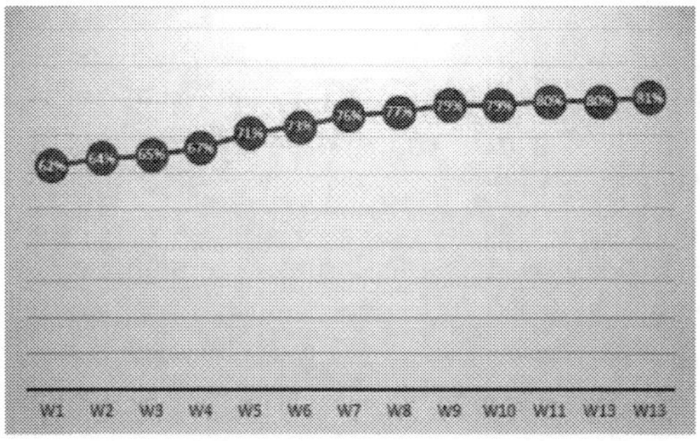

Figure 2. Percent Portfolio Projects Completed

Resource leveling is a technique which aims at completing projects adhering to their due dates within a given resource usage. That is to level all resource usages as much as possible throughout the total project durations. Based on the result of the critical path method, the resource leveling is defined based on the project activity schedules without violating precedence constraints (Figs. 3 & 4).

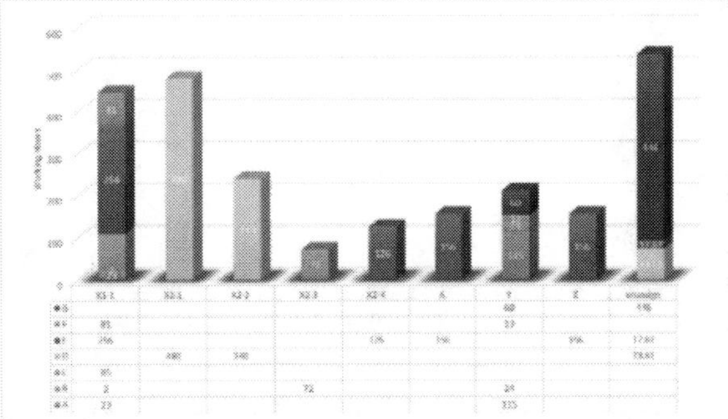

Figure 3. Resource hours before resource leveling

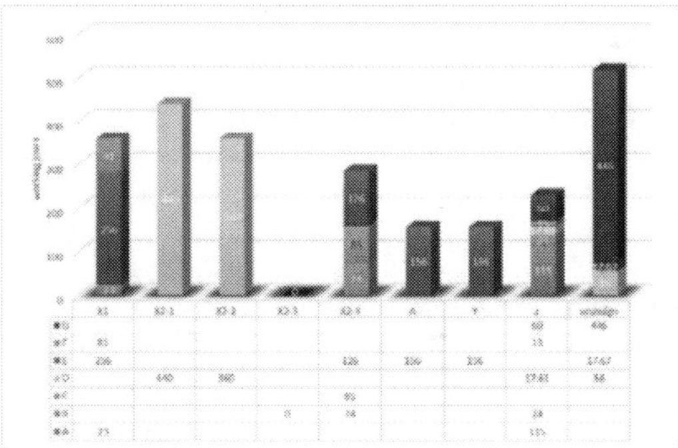

Figure 4. Resource hours after resource leveling

3.2 Portfolio Communication Management

The communication schedule consists of three communication areas reported monthly, bi-weekly, and weekly, depending on the requirements and details of communication (Table 1).

Table 1. Communication schedule

Communication Area	Frequency	Intended Recipient	Communication Vehicle
Portfolio Governance	Monthly	SEA Director, SEA PPM, Resource Manager, SEA Business Manager, Country PPM, Country Project Manager, Resources, Co-Project	1. Monthly Meeting 2. Monthly Report in internal Portal

Portfolio Performance Reports, Key Risks and Issues Updates	Bi-Weekly	Country PPM, Country Project Manager, Resources, Co-Project	1. Bi-Weekly Meeting 2. Bi-Weekly Report in internal Portal
Portfolio Dashboard	Bi-Weekly	SEA Business Manager Country PPM, Co Project	1. Bi-Weekly Meeting 2. Updated Performance Report in internal Portal 3. Email Distribution
Resource Utilization	Weekly	Resource Manager, Country PPM, Country Project Manager	1. Monthly Meeting 2. Resource Utilization Update Report 3. Email distribution

For a monthly report, based on a meeting report, each country reports its portfolio project status, key risks, issues, and required resources informed by the resource managers. For a bi-weekly report, the report is written after a local country meeting is completed. The country's project portfolio manager leads the meeting raises updated key risks and issues everyone in the project team needs to know in the meeting, and follows up on issues raised in the previous meeting. The last communication vehicle is resource utilization in which the resource manager leads the meeting to get issues from the country's engagement manager, project manager, and consultants who support the country's projects.

3.3 Standard Process Checklist

Three main processes are indicated in the project management process, i.e. (1) initial, (2) performance and (3) closure. All standard processes created by multiple teams are then listed and combined into a single checklist to be set as a standard for all teams and all team members.

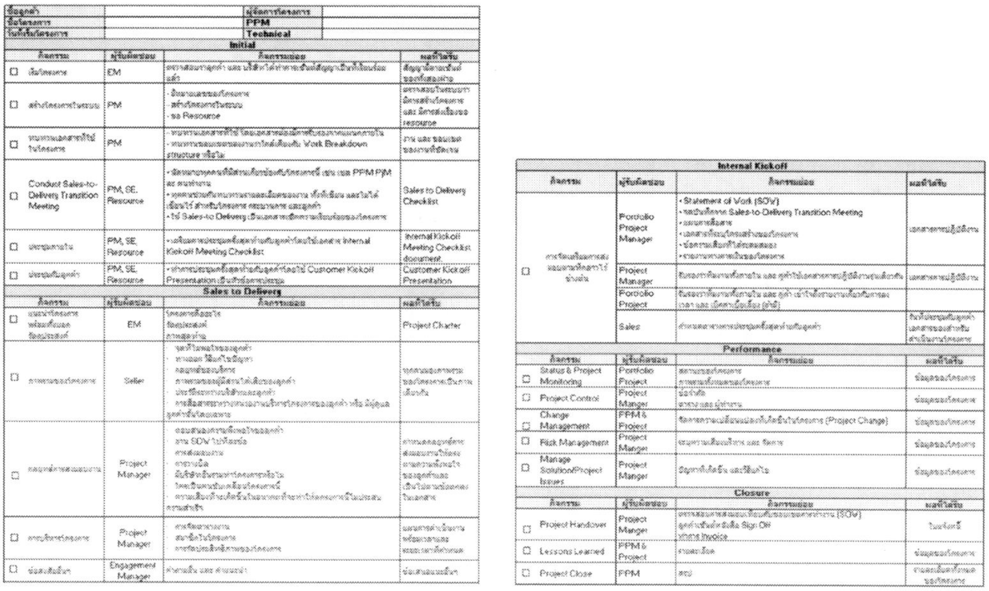

Figure 5. Checklist for Project Management Standard Process

4. Conclusion

The main problem of this case study which impacts delay-delivery is the lacking of project monitoring and tracking. This is due to project teams having to handle multiple projects simultaneously without sufficient skills in multiple tasks management. Relating to the objective of this research, multiple projects in an information technology consulting company needs to be improved to achieve higher percentage of on-time delivery. This can be implemented through the adapted techniques of project portfolio management. The theories of portfolio project management, project management information system, and project communication management are applied to develop appropriate methods to solve the problem.

After PMIS is employed to the case study, it results in the reduction of the pending jobs and provides the ability to track percent completion of portfolio projects. In addition, the percent completion of activities in each project can be easily tracked. The resource leveling can be applied to multiple consultant resources by the assistance of Microsoft Project.

Communication schedule is created as a standard communication process and sent recurrence meeting to those concerns, from top management to operation levels. Continuous notification is given to all team members to acknowledge the status and performance of ongoing projects including new coming projects to identify key risks, special issues, and resource requirements for all projects in South East Asia. This two-way communication helps the regional manager to allocate available resources to each country.

Finally, a standard guideline is developed as a regulation for sellers and project teams to follow which can reduce unsatisfactory steps and over scope requirements from customers. The main benefit is to avoid miscommunication and information distortion while transferring from customers to sellers and then to project teams.

5. References

[1] IEEE Guide - Adoption of the Project Management Institute 2011 Standard A Guide to the Project Management Body of Knowledge Fourth Edition
[2] Project Management Institute 2013 the Standard for Portfolio Management Third Edition
[3] Satawin, R. 2011, Project Portfolio Management System
[4] Marjolein C.J., Ralph, C. and Bakens, J.J.M. 2012 The effects of Project Management Information Systems on decision making in a multi project environment International Journal of Project Management 30 162-175
[5] Lee, S-K. and Yu, J-H. 2012 Success model of project management information system in construction Success model of project management information system in construction 25 82-93
[6] Turner, J.R. 2004 Communication and Co-operation on Projects between the Project Owner as Principal and the Project Manager as Agent European Management Journal 22 327–336
[7] Ochieng, E.G. and Price, A.D.F. 2010 managing cross-cultural communication in multicultural construction project teams: The case of Kenya and UK
[8] Braglia, M. and Frosolini, M. 2012 an integrated approach to implement Project Management Information Systems within the Extended Enterprise International Journal of Project Management 28 449 - 460
[9] Masmoudi, M. and Haït, A. 2013 Project scheduling under uncertainty using fuzzy modelling and solving techniques, Engineering Applications of Artificial Intelligence 26 135–149
[10] Biafore, B. 2013, Microsoft Project 2013: The Missing Manual, 515 - 565

Repeatability and Reproducibility of Fibre-Based Nanogenerator Synthesized by Electrospinning Machine

Suyitno[1,2], Sholiehul Huda[1], Zainal Arifin[1,2], Syamsul Hadi[1,2], Raymundus Lullus Lambang[1,2]

[1]Mechanical Engineering Department, Sebelas Maret University, Jl. Ir. Sutami 36 A, Surakarta, Central Java, Indonesia
[2]Research Center for Applied Energy Conversion and Nanotechnology, Sebelas Maret University, Jl. Ir. Sutami 36 A, Surakarta, Central Java, Indonesia

E-mail: suyitno@uns.ac.id, sholiehulhuda@gmail.com, zainal_a@uns.ac.id, syamsulhadi@ft.uns.ac.id, lulus_l@yahoo.com

Abstract. Zinc oxide fibres-based nanogenerators synthesized easily by electrospinning machine are promising to harvest electricity from mechanical energy. However, the repeatability and reproducibility were two major factors needed to be investigated to minimize product failure and to determine the feasibility of mass production of nanogenerators. The green fibres of zinc oxide were produced by electrospinning machine of zinc acetate and polyvinyl alcohol solution at a flow rate of 4 μL/min followed by sintering at temperature 550°C with heating rate 240°C/h. Each 10 nanogenerators was tested by three trained operators with three times of repetition at compressive load 0.5 kg. The nanogenerators revealed the maximum output voltage ranging from 203 to 217 mV. The value of repeatability and reproducibility of nanogenerators was approximately 24.29% showing that nanogenerators were still acceptable to be mass-produced. The relatively low reproducibility was mainly due to the operators, so that the checklist needed to be made easier and simpler for all the variables affecting to the quality of the fibres. Reducing the value of the repeatability and reproducibility is interesting to study further by creating a rotating collector so that the thickness and orientation of fibres can be arranged better.

1. Introduction

The scarcity of fossil energy sources and the increase of energy consumption worldwide have attracted engineers to develop renewable energy sources. Nanogenerator (NG) is one of energy converter devices that convert mechanical or thermal energy as produced by a small-scale physical change into electricity. Piezoelectric materials can be used as NG in which they convert mechanical energy into electrical energy by utilizing crystalline materials with non-centro symmetric. Moreover, NG is the technology of compact portable electric power sources that is widely developed at this time [1].

Some piezoelectric materials have been explored such as: PZT [2], $BaTiO_3$ [3, 4], PVDF [5, 6], and ZnO [7, 8]. The highest voltage generated from the PZT material, PVDF, and PVDF-TrFE was 1.63

[1] Mechanical Engineering Department, Sebelas Maret University, Jl. Ir. Sutami 36 A, Surakarta, Central Java, Indonesia, 57126. Phone/Fax: +62 (271) 632163. E-mail: suyitno@uns.ac.id.

V, 2.21 V, and 400 mV, respectively [9]. In contrast, ZnO-based NGs were more resistance at high temperature. In addition, ZnO-based fibers NGs generated DC power that did not require a rectifier. The DC type of NGs was very suitable to build a self-powered system that operated independently and wirelessly without the use of external energy storage systems. Furthermore, to obtain continuous electrical energy and large energy density, then the NGs were fabricated based on nanofibers synthesized by electrospinning machine. Nanofibers via electrospinning have been widely used because of their ease of the fabrication process and low cost [10-12].

Moreover, the piezo effect-based NGs were influenced by several factors: size, shape, type, and crystallinity of the materials. In addition to the raw materials, there were some other important parameters in the fabrication process of nanofibers with an electrospinning machine, including voltage, collector type, collector temperature, tip distance, needle diameter, needle angle, and precursor flow rate [13]. Naturally, nanofibers from electrospinning machine were randomly oriented, so that the output voltages generated by the piezo effect-based NGs were vary. Therefore, the repeatability and reproducibility aspects for fabricating nanofibers by electrospinning method are extremely needed for the purposes of mass production. However, the studies of the repeatability and reproducibility of producing NG based on nanofibers are still limited, especially for the nano-fibres of zinc oxide (ZnO).

2. Materials and Methods

2.1. Preparation of precursor materials

Polymer used was polyvinyl alcohol (PVA, MW = 70,000, Merck Ltd., Germany). The interesting properties of PVA polymer were soluble in water and some types of salts, especially zinc acetate and fully decomposed at a temperature of 400–500°C temperature [14]. To prepare the precursor materials, first, *the PVA solution* was made from 2 g of polyvinyl alcohol and 20 ml of distilled water. The solution was stirred at 70°C for 4 h [15] and allowed to settle at room temperature for 8 h. To produce the *ZnAc solution***,** 2 grams of zinc acetate dihydrate $Zn(CH_3COO)_2.2H_2O$ (Merck Ltd., Germany) was mixed with 8 ml of water and stirred for 1 h at 70°C. ZnAc solutions were then mixed with the PVA solution at a weight ratio of 1:4, stirred at 70°C for 8 h, and settled at room temperature for 24 h. This produced the ZnAc/PVA solutions and then ready for electrospinning.

2.2. The fabrication of fibers by electrospinning machine

The electrospinning machine was employed to produce green fibres. The ZnAc/PVA solution of 1 mL was put in the syringe pump of the electrospinning machine. The needle in the electrospinning machine was horizontally connected to the positive terminal to inhibit the beads on the collector while the aluminium plate was connected to the negative terminal. The distance between the terminals is 8 cm. A voltage of 15 kV was applied to push out the solution at a flow rate of 4 μL/min. The fibres were attracted by the electrostatic field and they attached themselves to the surface of the negative plate collector as green fibres. The green nanofibers were then sintered performed at a temperature of 550°C with a heating rate 240°C/h [16] to remove PVA and acid content of acetate and remained ZnO nanofibers. The morphology of the ZnO nanofibers was studied by scanning electron microscopy (SEM, FEI: Inspect-S50)

2.3. The fabrication of NGs

The NGs were made of aluminium plate-ZnO nanofiber-and aluminium plate as shown in Figure 1. Between the two surfaces of aluminium, plates facing each other were glued by PDMS to prevent a short circuit.

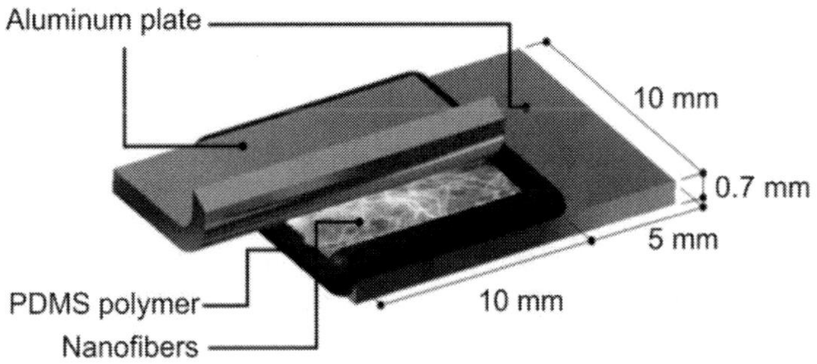

Figure 1. Schematic of NG

2.4. NGs testing

After completion for synthesizing ZnO fibres, piezoelectric materials were ready to be tested piezoelectricity, repeatability, and reproducibility with statistical sampling techniques. The performance of NGs was tested by direct measurement to determine the output voltage (V) with time (t) during the excitation process [17]. The output voltage was recorded by real-time data-acquisition hardware using Advantech USB-4716 and PCLS-Adam View 32 software. The NG was subjected to a compressive load (loading and unloading) of 0.5 kg for 3 s per cycle representing the pressure load of weak finger-taps [18]. Each 10 NGs were tested by three trained operators with three times of repetition [19]. The maximum voltage results were then calculated to determine the percentage of the repeatability and reproducibility (R&R) based on UCL (Upper Control Limits) and LCL (lower control limits) [20].

$$UCL_{\bar{\bar{X}}} = \bar{\bar{X}} + A_2\bar{\bar{R}} \tag{1}$$

$$LCL_{\bar{\bar{X}}} = \bar{\bar{X}} - A_2\bar{\bar{R}} \tag{2}$$

Where $\bar{\bar{X}}$ is the average of the sample means. $\bar{\bar{R}}$ is the average range of samples. A_2 is 0.31 for the sample size of 10.

The measure of the R&R was calculated by the following formulas.

Repeatability (EV):

$$EV = 5.1k_1\bar{\bar{R}} \text{ or } \%EV = 100\left(\frac{EV}{TT}\right) \tag{3}$$

Where k_1 is 0.5908. Next, reproducibility (AV) was calculated by the following:

$$AV = 5.1\left((k_2\bar{\bar{X}}_{diff})^2 - (EV^2/nr)\right)^{1/2} \text{ or } \%AV = 100\left(\frac{AV}{TT}\right) \tag{4}$$

Where k_2 is 0.5231. Furthermore, repeatability & reproducibility ($R\&R$) was determined by:

$$R\&R = (EV^2 + AV^2)^{1/2} \text{ or } \%R\&R = 100\left(\frac{R\&R}{TT}\right) \tag{5}$$

Meanwhile, the part of variation (PV) was:

$$PV = 5.1k_3R_p \text{ or } \%PV = 100\left(\frac{PV}{TT}\right) \tag{6}$$

Total tolerance was determined by the following:

$$TT = \text{upper spec} - \text{lower spec} \tag{7}$$

Where k_3 is 0.3146.

3. Results and Discussions

3.1. Output voltage from NGs

The study to determine the repeatability and reproducibility of piezoelectric-based NGs for producing energy was conducted by measuring the output voltage in real-time. The NGs were subjected to the compressive load of 0.5 kg. The sample of the output voltage profile of ZnO fibres-based NGs is shown in Figure 2. The piezo effect occurred because the pressure applied by the external load was transferred into nanofibers. Furthermore, electrons generated in the nanofibers were then transferred through the electrode [2].

Figure 2 shows also the steady decline of the output voltage as time passes. This is because at the beginning of loading, the piezo materials in NGs had the intact fibre networks. Further, the applied load caused the fibre networks gradually experiencing the compaction. Over time, some fibre structures had torn as shown in Figure 3. When the fibre networks decayed, the NGs showed the drop of voltage [21]. Despite the damage to the fibre networks, the piezoelectric-based NGs might still produce output voltage because the crystalline structure was not broken due to the damage of the fibre networks.

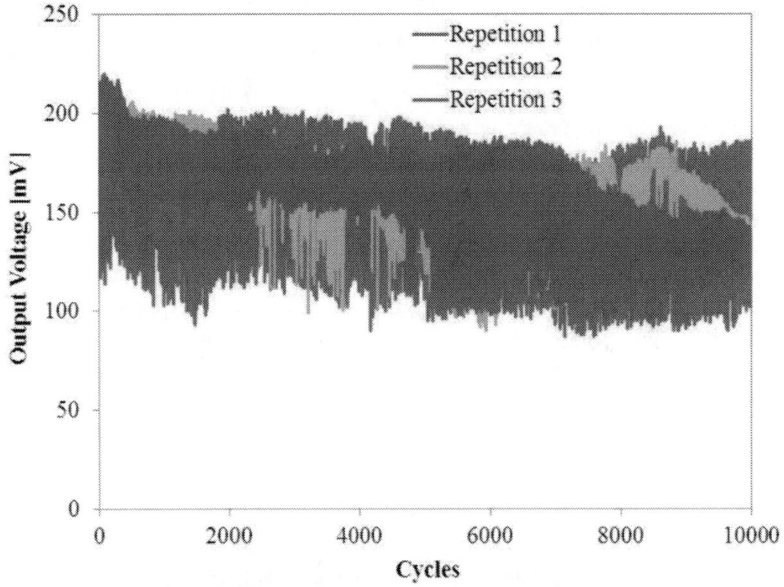

Figure 2. Sampling output voltage from NGs

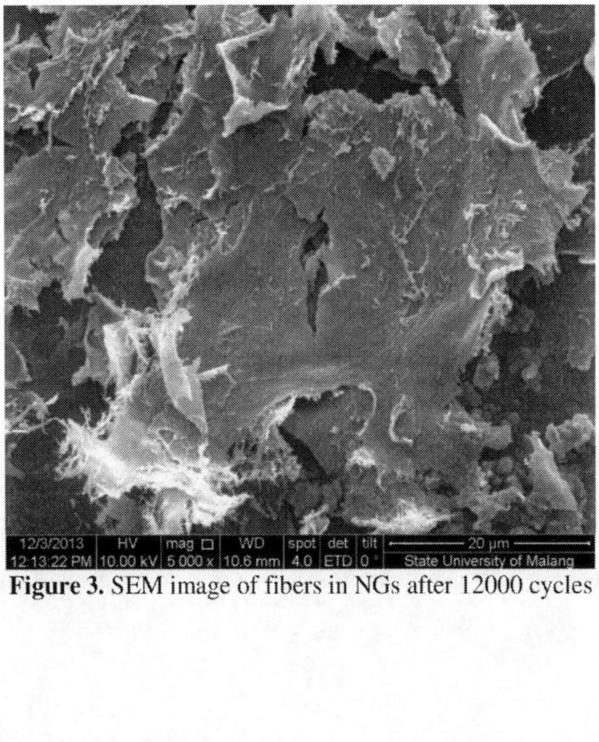

Figure 3. SEM image of fibers in NGs after 12000 cycles

In addition, the piezo material-based NGs had a recovery time phenomenon that was the time required to stand back into their initial state. The release of a load (unloading) was part of the recovery process of nano-fibers aimed to maintain the stability of the excitation for the piezoelectric-based NGs to produce the output voltage.

3.2. Repeatability and reproducibility (R&R) of NGs.

The analysis of R&R for NGs was conducted on the value of the maximum output voltage of the NG (see Table 1). The highest and lowest maximum output voltages of 10 samples tested by three operators with three repetitions were 217 mV and 203 mV, respectively. The maximum voltage indicated the performance of NGs as an electric energy harvester from mechanical energy. Furthermore, a large voltage was required when the NGs would be used as an electric charger. However, when viewed from the production process, the magnitude of the maximum voltage was varying for each part; therefore, it was very necessary to measure the repeatability and reproducibility of the synthesis process of NGs.

Table 1. The maximum output voltage produced by NGs

Appraiser/ Trial #		Parts									
		1	2	3	4	5	6	7	8	9	10
	1	214	214	211	209	215	209	214	215	215	216
A	**2**	208	214	208	216	214	214	216	216	207	216
	3	212	206	216	216	208	215	215	216	214	213
	1	210	215	213	215	215	215	215	214	214	214
B	**2**	214	212	209	215	207	211	213	208	207	216
	3	211	208	208	207	207	214	209	214	211	216
	1	203	210	203	207	212	204	206	204	205	203
C	**2**	206	213	204	203	203	205	204	206	205	203
	3	203	215	212	214	203	206	217	205	212	203

The study revealed that the repeatability, the reproducibility, and the repeatability and reproducibility (R&R) based on total tolerance were 17.48%, 16.86%, and 24.29%. According to quality control principles [22], if R&R is below 10%, the process is acceptable. If R&R is between 10 and 30%, the process is then acceptable based on the importance of the application and associated costs. If R&R is more than 30%, then the process is considered unacceptable and therefore cannot be made in mass production. Thereby, the NGs demonstrated the greater repeatability (16.86%) than the reproducibility (24.29%). Based on total tolerance, the effect of the equipment, particularly an electrospinning machine to produce uniform NGs was bigger than that of the operators. The electrospinning machine produced random fibres [13] attracted by the electrostatic field generated by the high voltage source. The forming process of fibres followed the direction of electrons movement from the needle tip to the collector. The greater the difference in the electrostatic field between the tip of the needle and the particular area of the collector, the thicker fibres were formed in the collector region. The collector area that has been met with thicker fibres caused the decrease of electrostatic field differences, and the following fibres moved to a different area of the collector having the larger electrostatic field. This process was difficult to control for a large enough area of the collector (5 x 5 cm). In addition, polymer jets coming out of the needle tip moving toward the collector was usually in a chaotic oscillating motion. The wider the collector caused the greater the chaotic oscillating motion leading to the more non-uniform distribution of fibers on the wide collector. Reducing the size of the collector might obtain the more uniform distribution of fibers on the collector. Thus, to study further to avoid this, it can be done by (1) creating a smaller collector area or (2) creating a moving or rotating collector so that the flatness of the fibres on the surface of the collector can be better.

Furthermore, the percentage of appraiser variation of 16.86% was acceptable [22]. Reproducibility was determined from the variability created by several operators measuring each part several times, effectively quantifying the variation in a measurement system resulted from the operators and environmental factors such as time [22]. The repeatability of 17.48% that was higher than the reproducibility of 16.86% indicated that the effect of part variation was greater than that of operator variation.

However, the performance of the first operator and the second operator was almost the same, but the performance of the third operator was still below the first operator and second operator as shown in Figure 4. The differences in the output voltage of the NGs produced had a wide range approximately 14 mV. The maximum voltage of NGs resulted was not stable when repeated by different operators. The existence of the operator factors caused the uniformity of ZnO fibres-based NGs have a relatively low value. For further study, the number of variables affecting the quality of the fibres-based NGs need to be trained to the operators, including what and how the influence of these variables on the quality of the fibres. The checklist of each variable and measuring the magnitude of each variable should be made easier and simpler so that the processing time is shorter. Furthermore, the inconsistency because of the operators can be minimized.

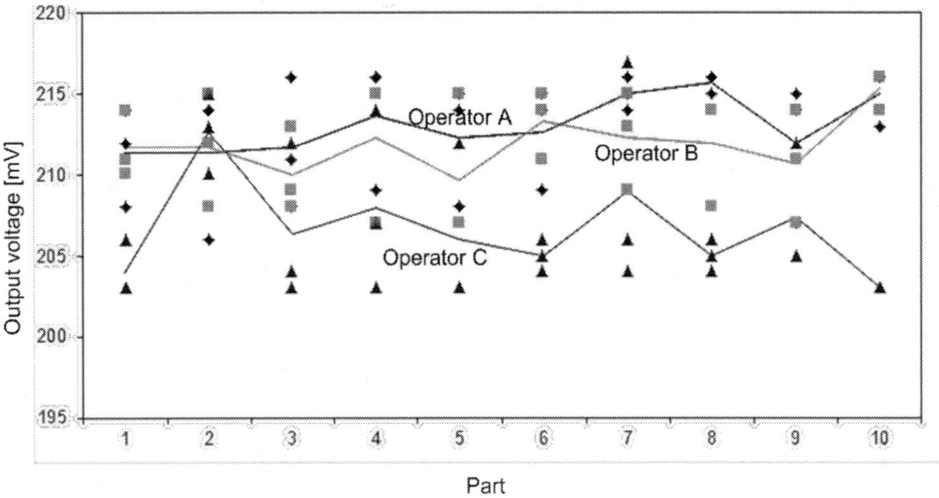

Figure 4. A control chart influenced by operators

4. Conclusion

ZnO fibres-based NGs have been successfully produced by electrospinning machine. The output voltages of NGs when subjected to the load of 0.5 kg were between 203 mV and 217 mV and suited to be developed as an electric charger. Over time, the voltage generated from ZnO fibres-based NGs declined caused by the damage to the fibrous structure. The R&R value of NGs at 24.29% was still acceptable so that NGs can be mass-produced. Furthermore, bringing down the value of the R&R of NGs can be done by (1) creating a smaller collector or (2) creating a rotating collector so that the thickness of the fibres can be arranged better. The relatively low reproducibility was due to the operators, therefore the checklist needed to be made easier and simpler for all the variables affecting to the quality of the fibres.

Acknowledgment

The authors thank the Rector of Sebelas Maret University for financial support through the UNS Research Grant 2013 and 2014.

References

[1] Fry D N, Holocomb D E, Munro J K, Oakes L C, Mason M J 1997 *Compact portable electric power sources* (USA: Oak Ridge National laboratory)

[2] Chen X, Xu S, Yao N, Shi Y 2010 1.6 V nanogenerator for mechanical energy harvesting using PZT nanofibers *Nano Lett.* **10** 2133–7

[3] Park K I, Xu S, Liu Y, Hwang G T, L K S J, Wang Z L, et al. 2010 Piezoelectric $BaTiO_3$ thin film nanogenerator on plastic substrates *Nano Lett.* **2**(12) 4939-43

[4] Terashi Y, Purwanto A, Wang W-N, Iskandar F, Okuyama K 2008 Role of urea addition in the preparation of tetragonal $BaTiO_3$ nanoparticles using flame-assisted spray pyrolysis *J Eur Ceram Soc.* **28** 2570-80

[5] Indolia A P, Gaur M S 2012 Investigation of structural and thermal characteristics of PVDF/ZnO nanocomposites *J Therm Anal Calorim.* **113**(2) 821-30

[6] Lee H, Cooper R, Wang K, Liang H 2008 Nano-scale characterization of a piezoelectric polymer (polyvinylidene difluoride, PVDF) *Sensors.* **8**(11) 7359-68

[7] Park J A, Moon J, Lee S J, Lim S C, Zyung T 2009 Fabrication and characterization of ZnO nanofibers by electrospinning *Current Applied Physics.* **9** 210-2

[8] Ren H, Ding Y, Jiang Y, Xu F, Long Z, Zhang P 2009 Synthesis and properties of ZnO nanofibers prepared by electrospinning *Sol-Gel Science Technology.* **4** 287-90

[9] Chang J, Dommer M, Chang C, Lin L 2012 Piezoelectric nanofibers for energy scavenging applications *Nano Energy.* **16** 356-71

[10] Dong Z, Kennedy S J, Wu Y 2011 Electrospinning materials for energy-related applications and devices *J Power Sources.* **196**(11) 4886-904

[11] He J H, Liu Y, Mo L F, Wan Y Q, Xu L 2008 *Electrospun nanofibres and their applications* (United Kingdom: iSmithers)

[12] Ren H, Ding Y, Jiang Y, Xu F, Long Z, Zhang P 2009 Synthesis and properties of ZnO nanofibers prepared by electrospinning *J Sol-Gel Sci Technol.* **52**(2) 287-90

[13] Pham Q P, Sharma U, Mikos A G 2006 Electrospinning of polymeric nanofibers for tissue engineering applications: A review *Tissue Engineering.* **12** 1197-211

[14] Yang X, Shao C, Guan H, Li X, Gong J 2004 Preparation and characterization of ZnO nanofibers by using electrospun PVA/zinc acetate composite fiber as precursor *Inorganic Chemistry Communications.* **7**(2) 176-8

[15] Park J Y, Kim S S 2009 Growth of nanograins in electrospun ZnO nanofibers *J Am Ceram Soc.* **8** 1691-4

[16] Lee D Y, Cho J-E, Cho N-I, Lee M-H, Lee S-J, Kim B-Y 2008 Characterization of electrospun aluminum-doped zinc oxide nanofibers *Thin Solid Films.* **517**(3) 1262-7

[17] Choi D, Lee K Y, Lee K H, Kim E S, Lee S Y, Kim S W, et al. 2010 Piezoelectric touch-sensitive flexible hybrid energy harvesting nanoarchitectures *NanoTechnology.* **21** 1-6

[18] Hsu H L, Yang C B, Huang C H, Hsu C Y 2013 Electrical and optical studies of Ga-doped ZnO thin films *J Mater Sci: Mater Electron.* **24** 13–9

[19] Besterfield D H 1994 *Quality control* (United States of America: Prentice Hall, inc.)

[20] Woodall W H, Adams B M, Benneyan J C 2011 *The use of control charts in healthcare* (USA: Department of Statistics. Virginia Tech.)

[21] Picu R C 2011 Mechanics of random fiber networks—a review *Soft Matter.* **7**(15) 6768

[22] Hoffa D W, Laux C 2007 Gauge r&r: An effective methodology for determining the adequacy of a new measurement system for micron-level metrology *Journal of Industrial Technology.* **23** 9

Improving Operational System Performance of Internet of Things (IoT) in Indonesia Telecomunication Company

M Dachyar, S A Risky

Industrial Engineering Department.
University of Indonesia, Depok 16424, Indonesia

E-mail: mdachyar@yahoo.com

Abstract. Telecommunications company have to improve their business performance despite of the increase customers every year. In Indonesia, the telecommunication company have provided best services, improving operational systems by designing a framework for operational systems of the Internet of Things (IoT) other name of Machine to Machine (M2M). This study was conducted with expert opinion which further processed by the Analytic Hierarchy Process (AHP) to obtain important factor for organizations operational systems, and the Interpretive Structural Modeling (ISM) to determine factors of organization which found drives the biggest power. This study resulted, the greatest weight of SLA & KPI handling problems. The M2M current dashboard and current M2M connectivity have power to affect other factors and has important function for M2M operations roomates system which can be effectively carried out.

1. Introduction

The current market value of IoT services in Indonesia has increased significantly since 2010. Increasing of IOT user, was certainly supported by mobile subscribers which in 2013 has reached 265 million users. The growth has pushed the interaction of machines as supporting information technology. Telecommunication companies can provide services in accordance to consumer, design a better system to support IoT operational performances. The structure of the report and information flow engaged in IoT, which part is considered more important and critical than others, producing maximum operating systems. With the framework of operational information systems in the IoT division will help the company to execute its operational systems synergistically and achieved successfully services for customers.

2. Literature Review

2.1 Internet of Things

Internet of Things as known by the acronym IoT, has evolved the need to interact to internet at anywhere. IoT is a concept that aims to add the benefits of internet connectivity as continuously connected. The ability of IoT as data sharing, as a remote control, can make connections to real-world objects. As for example; foodstuffs, electronics, collectibles, the equipment can all be connected to local and global networks via embedded sensors and always active. Currently IoT concept has been used in entire body with a network-based RFID technology [1].

2.2 Operation System Performance IoT on Telecommunications Company

Performance of company depends on services that has to be provided by company. Currently IoT operating system performance in telecommunications are not run in synergy which mutually not coordinated. The IoT operational system were done individually without coordination to each system, so the services can not meet the consumer needs. To improve the performance of company, the IoT coordination of all operational systems must be taken as to increase customer satisfaction.

In providing the best service to consumers, every company must have operating systems that support ministry. The IoT chart of operational systems has been established to support IoT services [2].

The IoT operating system consists of several criterias and sub-criterias [3]: technical dashboard – helpdesk, technical dashboard – application layer, technical dashboard – connectivity layer, customer complaint or problem handling, infrastructure or facility, NOC as services, training, M2M customer, and technical dashboard – network / system legacy.

2.3 Multi Criteria Decision Making (MCDM)

MCDM methods are aimed for the decision-making criteria, mutually conflict, and has sizes that can not be compared to each other. MCDM solve problems in quantitative or qualitative criteria [4]. AHP is a decision-making simple method in handling complex problems, with lots of unstructured attributes. AHP analysis models with orderly facing complexity problems using hierarchical structure forming a multi-criteria decision-making approach uses pairwise comparisons [5]

2.4 Interpretive Structural Modeling (ISM)

ISM is an in interactive learning process. In this technique, a set of differences are structured into a comprehensive systematic model. The model portrays structure of a complex issue or problem in a designed pattern implying graphics as well as words [6].

Interpretive Structural Modeling (ISM) is a well-established methodology for identifying relationships between IoT operational factors. Direct and indirect relationships operational factors describe the situation far more accurately than the individual factor taken into isolation [7].

This methodology is primarily intended as a group learning process, but individuals can also use it. The various steps involved in the ISM technique are as follows: i. Identify the elements which are relevant to the problem. ii. Establish a contextual relationship between elements with respect to which pairs of elements would be examined. iii. Develop a structural self-interaction matrix (SSIM) of elements. This matrix indicates the pair-wise relationship among elements of the system. This matrix is checked for transivity. iv. Develop a reachability matrix from the SSIM. v. Partition the reachability matrix into different levels. vi. Convert the reachability matrix into conical form. vii. Draw digraph based on the relationship given in reachability matrix. viii. Change resultant digraph to ISM-based model by replacing element nodes with the statements. ix. Review the model to check for conceptual inconsistency and make the necessary modifications [6]. Structure of self-interaction matrix (SSIM); ISM methodology suggest the use of the expert opinions based in various management techniques such as brain storming and NGT.

3. Research Methodology

This study has uses AHP method, through expert interviews to judge the important factor.Expert judging factors on pairwise matrix calculation. The results of calculations resulting factor matrix pairs with weights. Factor with value weights are used as input to the organization method Interpretive Structural Modeling (ISM) to generate structural models of organizational information flow. Expert interviews and through brainstorming has assessed the relationship between factors to produce contextual relationship between elements to develop a structural self-interaction matrix (SSIM).

4. Result and discussion

Questionnaire with a scale of 5 was given to four (4) experts to gain organizational elements which are important in the IoT pada metode AHP. The results of the questionnaire were processed and searched Geomin values above 3.5, obtained no criteria 9 and 24 factors. Expert judging factors on pairwise matrix calculation. The results of calculations resulting factor matrix pairs with weights Subsequent questionnaires given to experts to get an assessment of the degree of importance of the criteria with other criteria. All processed calculated by the AHP Expert Choice software. Produced the most important factor is the weight of SLA & KPI problem handling 0.275 (Figure 1).

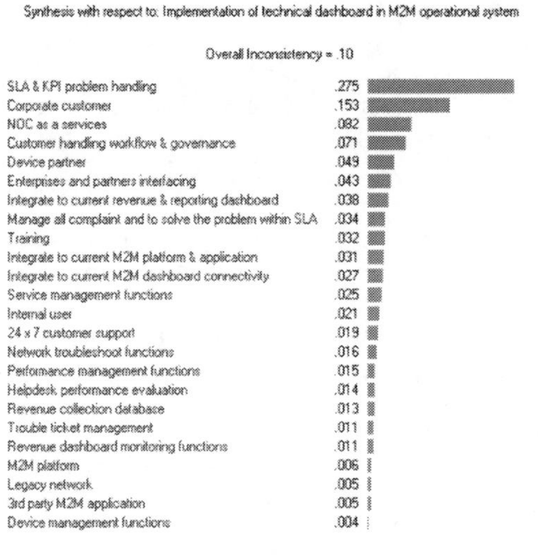

Figure 1 The weight of each factor

The next stage is using the Interpretive Structural Modeling (ISM) method and to create framework model. There are 24 factors established from expert judment from previous method. A contextual relationship between factors V: $e_{ij} = 1$; $e_{ji} = 0$, A: $e_{ij} = 0$; $e_{ji} = 1$, X: $e_{ij} = 1$; $e_{ji} = 1$, O: $e_{ij} = 0$; $e_{ji} = 0$. Expert has assessed the correlation of between 1 and 24 is v, assess the correlation between 1 and 23 is A, further illustrated in Structural Self-Interaction Matrix (Figure 2).

VARIABEL	24	23	22	21	20	19	18	17	16	15	14	13	12	11	10	9	8	7	6	5	4	3	2	1
1	V	A	A	A	V	X	X	A	X	X	X	A	A	A	X	X	X	X	X	X	X	X	X	
2	O	A	A	O	O	X	X	A	A	A	A	A	A	A	A	A	A	A	A	A	A	A		
3	A	A	A	A	V	V	V	A	A	A	A	A	A	A	A	X	A	A	A	A	V			
4	A	A	A	A	V	V	V	O	A	A	A	A	A	A	A	V	A	A	A	A				
5	A	A	A	A	V	V	V	O	X	X	A	A	A	A	V	V	X	V	V					
6	O	A	A	O	V	V	V	V	V	V	V	O	A	A	O	O	V	O						
7	A	A	A	A	V	V	V	O	O	O	V	A	A	A	A	O	X							
8	A	A	A	A	V	V	V	A	A	A	A	A	A	A	A	V								
9	O	X	O	A	V	V	V	O	A	A	A	A	A	A	O									
10	X	O	O	A	V	V	V	O	X	X	V	X	X	X										
11	O	A	A	A	V	V	V	V	V	V	V	O	O											
12	V	V	V	V	V	V	V	V	V	V	V	O												
13	V	V	V	V	V	V	V	V	V	V	V													
14	A	A	A	V	V	V	V	X	X	X														
15	X	X	V	A	V	V	O	A	X															
16	A	X	X	A	V	V	V	A																
17	X	X	X	A	V	V	V																	
18	A	A	A	V	O	X																		
19	O	O	X	V	V																			
20	A	A	A	A																				
21	X	X	X																					
22	O	O																						
23	V																							

Figure 2 Structural Self-Interaction Matrix (SSIM)

Based on the result of SSIM, next is the reachability matrix processing as the second step of the ISM methodology. ISM framework resulted from the partition 9 times iteration, iteration to-1 the 20 variables as internal factor, have values equal to the intersection reachability sets. The next 20 variables are at level 1 and will eliminate the next iteration, and proceed to the next iteration. Through drawing Chronicle matrix, and last stage is structural digraph Including links transitivity, resulting

ISM framework M2M on operational systems (Figure 3). ISM framework shows how the level of interest in the relationship between organizational factors IOT starting IX level up to level I result in maximum work performance.

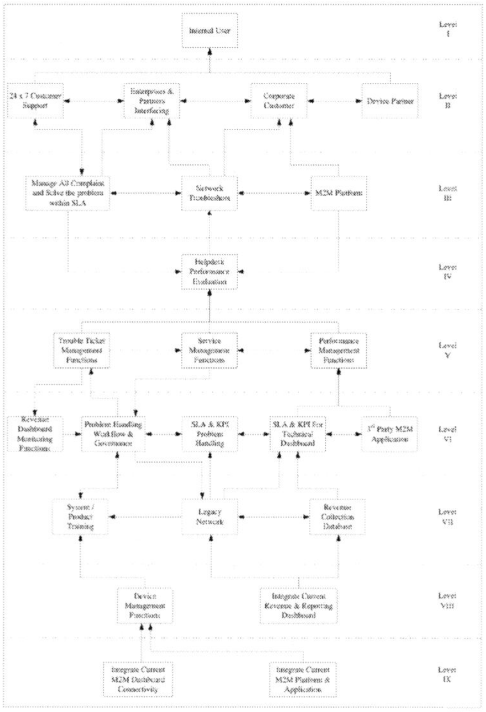

Figure 3 ISM framework on operational systems M2M

The level IX, has integrated M2M current dashboard, current M2M connectivity, and application platform which are governed a contextual relationship to other factors, and vice versa. This factors have power to affect other factors and has important function for M2M operations system which can be carried out effectively.

AHP and ISM Analysis

Weighting of the sub-criteria of global M2M operating system, shows it is not necessarily the highest weight by AHP, otherwise it has high power drives in the ISM . As can be seen, the factors at level 9, is the most influential factors in ISM model to integrate the current M2M platform with weight of 0,031 and integrate to current M2M connectivity dashboard with weight of 0.027.

Factors with the highest weight is the SLA & KPI handling problems with weight of 0.275, this variable at the level of VI, linkage has drives power and high dependence. This factors should carefully be studied, because of its strong driving force, effect, and interaction which can provide feedback to the system .

5. Conclusion

IOP Conf. Series: Materials Science and Engineering **58** (2014) 012014 doi:10.1088/1757-899X/58/1/012014

According to research objectives, the results of weighting the criteria and sub-criteria by AHP provides an overview to the company to focus sharply to problem of handling the SLA & KPI of 0.275, then corporate customers amounted to 0.153, and the problem of handling workflows and governance of 0.082. Second, retrieved draft framework of operational systems M2M with Interpretive Structural Modeling method (ISM) carefully describe the contextual information and organizational relationship between the operational factors in the M2M system. The initial framework design is to monitor the performance of the company's operating system so it can work synergistically and measurable. Based on the resulting framework, the factors influential to other factors in the height of the operational system is integrated to current M2M connectivity dashboards and integrate to current M2M platforms application.

References

[1] Ma, H.-D. (2011). Internet of Things: Objectives and Scientific Challenges. *Journal of Computer Science and Technology*, 26(6), 919-924.
[2] Yun, Miao (2010). *Research on the architecture and key technology of internet of things (IoT) applied on smart grid.* International Conference on Advances in Energy Engineering, Beijing
[3] ITU (2005,November). *ITU Internet reports 2005: the Internet of things.* December 3, 2013. http://www.itu.int/internetofthings/
[4] Chai, J. L. (2013). Application of decision making techniques in supplier selection: a systematic review of literature. *Expert Systems with Application*, 40(10), 3872-3885.
[5] Saaty, T. L. (2008). Decision making with the analytic hierarchy process. *International Journal Services Sciences*, 1(1), 83 - 98.
[6] Attri, R. D. (2012). An ISM approach for modelling the enablers in the implementation of total productive maintenance (TPM). *International Journal Assurance Engineering and Management*, 4(4), 313 - 326.
[7] Thakkar, J. D. (2007). An integrated approach of interpretive structural modeling (ISM) and analytic network process (ANP). *International Journal of Productivity and Perfromance Management*, 56(1), 25-59.

Angiogenesis in tissue engineering: from concept to the vascularization of scaffold construct

Siti Amirah Ishak[1], J.R. Pangestu Djuansjah[2], M. R. Abdul Kadir[1], Irza Sukmana[1, 3*]

[1]IJN-UTM Cardiovascular Engineering Centre (CEC) and Medical Devices and Technology Group (MEDITEG), Faculty of Biosciences and Medical Engineering, UTM Skudai, Johor, 81310 Malaysia, [2]Department of Mechanical Engineering, Universiti Teknologi Malaysia, UTM Skudai, Johor, 81310 Malaysia, [3]Department of Mechanical Engineering, University of Lampung, Jalan Prof. Soemantri Brojonegoro No. 1, Bandar Lampung 35143, Indonesia.

***Corresponding email:** irza.sukmana@gmail.com

Abstract. Angiogenesis, the formation of micro-vascular network from the pre-existing vascular vessels, has been studied in the connection to the normal developmental process as well as numerous diseases. In tissue engineering research, angiogenesis is also essential to promote micro-vascular network inside engineered tissue constructs, mimicking a functional blood vessel in vivo. Micro-vascular network can be used to maintain adequate tissue oxygenation, nutrient transfer and waste removal. One of the problems faced by angiogenesis researchers is to find suitable in vitro assays and methods for assessing the effect of regulators on angiogenesis and micro-vessel formation. The assay would be reliable and repeatable with easily quantifiable with physiologically relevant. This review aims to highlights recent advanced and future challenges in developing and using an in vitro angiogenesis assay for the application on biomedical and tissue engineering research.

1. Introduction

The word of angiogenesis comes from two Greek words, they are: "angêion" which means vessel and "genesis" means birth or emergence. Angiogenesis can also be defined as hemangiogenesis which means the formation of new blood vessels or lymphangiogenesis means the formation of new lymph vessels from the existing ones. In 1971, the "father of angiogenesis", Judah Folkman hypothesized that angiogenesis was a factor enabling malignant tumour growth in cancer [1].

Recently, there have been growing interests in the development of biomaterial for biomedical, especially in the field of tissue engineering. Tissue engineering has developed as new, interdisciplinary [2] and multidisciplinary research field by which it involves the use of living cells and developing a biological substitute for implantation into the human body. Besides that, biomaterial science and tissue engineering gives a great demand for the replacement of damaged tissue in this growing population [3]. These medical concepts represent a new era in our efforts to treat problems associated with failing tissues and organs.

Further advancements in tissue engineering also have brought significant knowledge on the mechanisms and parameters related to vascularization and angiogenesis development [4-5]. Tissue engineering scientific society relies on the increasing knowledge of vasculo- and angio-genesis within the polymer scaffolds. This review aims to contribute on the current status and future challenges on development of an assay to study and follow-up capillary formation and micro-vessels network promotion. Tissue engineering societies seeks knowledge on angiogenesis assay for further application on the vascularisation of thick engineered scaffold construct.

2. Angiogenesis assay

Over the past decades, 2D system has been converted into 3D system in order to ensure a higher physiological relevance of experiments. However, the limitations of 2D system in cell culture become increasingly evident [6-7]. While 3D culture systems are widely used in tissue engineering and biomedical research, 2D culture system is still entrenched in the pharmaceutical industry. 2D models refer to those in which cells develop a tubular structure on the surface of substrate meanwhile, 3D models refer to the cell that invade the surrounding matrix that consists of biogel such as collagen, fibrin and matrigel in order to form a 3D structures [8-9]. 3D system is much better than 2D system to stimulate in vivo environment because 3D system able to describe more steps involved in the angiogenesis process [10-11]. Furthermore, 3D system can be used to study the effect of gradient diffusion of nutrient and oxygen, accumulation of waste and stimulating factors on the event that occur during angiogenesis. Although 2D system is less expensive compared to 3D system but still 2D system cannot provide a third dimension that will correctly mimic in vivo surrounding [12]. 2D system also fails to express cell to cell communication [12], tissue- specific architecture and their environment lack of mechanical and biochemical signal. Based on this reason, this is why result obtained from 2D system is not very reliable because they are unable to mimic in vivo environment which is very important to study the angiogenesis process.

Angiogenesis is the process of new blood vessel formation from the existing vasculature [7-8]. Angiogenesis play an important role in the physiological condition such as wound healing. However, one of the problems faced by many researchers is to find suitable assays in order to assess the effect of regulators of the angiogenic response. The selection of appropriate assay neither in vivo nor in vitro has become the most important technical challenge to the studies of angiogenesis. In vitro assays are very useful to study the various steps occurring in the angiogenesis such as cell proliferation, cell migration, and cell differentiation [13]. Usually cell will undergo proliferation within the process of angiogenesis. Cell proliferation assay are commonly used because this assays are easily to perform and highly reproducible [14].

Cell proliferation is defined as the number of cell divide. In this assays, increased in cell number is measured by using haemocytometer. Another in vitro assay is cell migration assay by which during angiogenesis, cells are stimulated to degrade the basement membrane and migrate toward the angiogenic factor. The most frequently used assay to assess the migration of cells is Boyden chamber. In this assays, cells are placed on top of the layer of filter and permit the cell to migrate towards the angiogenic factor that placed in the lower chamber [14]. Finally, cell differentiation is the last assays that will stimulate the formation of tubules. This assays involved plating cells onto the layer of gel matrix consists of collagen, fibrin or Matrigel where it will stimulate the proliferation, attachment, migration and differentiation of the cells into tubules so that it can mirror the in vivo environment

[13]. In other word, this assay can be used to study the ability of cells to form a three- dimensional structure. Besides that, the rate of cell differentiation depends on the type of matrix used [15].

Angiogenesis also have been studied in vivo which has been initially developed to become more reliable and easy method that can permit better quantification. In vivo assays include chick chorioallantoic membrane (CAM) assay, corneal angiogenesis assay and many more. CAM assays are the most popular that have been used for in vivo assay. This is because, CAM assays is relatively simple and inexpensive for a large- scale screening. However, the limitation of CAM assays is that CAM has a well developed network, thus making it hard to distinguish new capillaries from the existing ones. Another assay known as corneal angiogenesis assay whereby this assay is reliable and easy to be monitored but expensive as compared to CAM assay therefore it is impractical for large-scale screening. In this assay, corneal is known as avascular tissue so any newly vessel formed, it can be readily visible and can be quantified. [16].

3. Extra-cellular matrix (ECM) for angiogenesis study

Angiogenesis is a dynamic process during physiological condition especially wound healing where a new blood capillary is developed and remodelled again to replace the damaged vessel. During angiogenesis process where migration of cell and development of new capillaries occurred, they will not only depend on the cell and cytokines present but also the production and organization of extracellular matrix component. Formation of a new blood capillary will also depend on the extracellular matrix component because ECM will not only serve as structural support during the developing vasculature but also providing some information guidance for new capillaries [17]. Besides that, the interaction between extracellular matrix and cell are very crucial and further required for complete the vessel maturation. Besides gives support to the cell and tissues, ECM also support the adhesion of cell and transduces cell signal [18]. In addition, ECM is mainly composed of two classes of macromolecules which are glycosaminoglycans and fibrous protein such as collagen, elastin, fibronectin and laminin [12]. Collagen is one of the major structural proteins in the ECM where many researchers have used it as a potential scaffold in vascular tissue engineering. However, there are some limitations of using collagen where it does not possess high mechanical strength which may result in a weak construction for implantation [19].

Besides collagen, fibrin is also known to be one of the native ECM component that involved angiogenesis. This is because; fibrin shows excellent properties as the matrix where it has been widely used to investigate the angiogenic behavior of different types of endothelial cells in vitro with the presence of growth factor [20]. Fibrin is one of the natural proteins that shows to support the vascular growth meanwhile another synthetic materials are effectively inhibit angiogenesis unless they are impregnated with growth factor [21].

4. Angiogenesis assessment and imaging

Angiogenesis is known as a basic process that involved in a variety of physiological as well as pathological condition. Physiologically, angiogenesis is essential for the wound healing, ischemia, and during development of embryo. Pathologically, it is associated with diseases such as cancer, blinding ocular disorders, rheumatoid arthritis and psoriasis. Besides that, the realization of angiogenesis as one of the essential process that potentially induce the growth of tumor, has led to enormous interest in the development of drugs that can target these molecules as well as discovery of angiogenesis stimulus. However, the difficulties in designing clinical trials for anti-angiogenic drugs causes some of the potential drugs can't live up to the expectation of preclinical trials [25]. Recently, imaging modalities may be a better alternative to assess and detect angiogenesis due to the fact that it is non-invasive and can be assess in a larger volume. Examples of modalities used are X-ray computed tomography (CT), magnetic resonance imaging (MRI), positron emission tomography (PET) and ultrasound.

MRI is one of the practical modality that has been used widely in assessing angiogenesis. MRI is a method used to estimate angiogenic blood flow and volume. Besides MRI, CT imaging is another modality that has been used to assess angiogenesis and shows more quantitative data than data obtained by MRI. Other than CT scan and MRI, ultrasound imaging also can be used to assess angiogenesis. Although ultrasound imaging is portable, widely available and inexpensive but it produce images that low resolution compared with those produces by CT imaging ad MRI. Another approach for the imaging of angiogenesis vasculature is the use of molecular markers. The example of molecular markers that have been identified including vascular endothelial growth factor receptor (VEGF), $\alpha_v\beta_3$ integrins, endoglin and endosialin [26]. These molecular markers will bind to the molecular agents to assess high specificity of angiogenesis epitope.

Currently, there are several experimental methods available to study on the quantification of angiogenesis activity instead of angiogenesis imaging. Method available include capillary density estimation, enzyme linked immunosorbent assay (ELISA), hemoglobin determination in matrigel plugs and semi-automated vessel quantification. Capillary density estimation is one of the methods that have been mostly used in the quantification of angiogenesis activity because this method is easy to perform in laboratory and less cost [27]. Besides that, enzyme linked immunosorbent assay (ELISA) is known as a test that used antibodies and color change to identify a substance. However, this method also can be used to estimate angiogenenic growth factors [28]. Another method known as matrigel vessel quantification in which this method is known as the simplest method to determine the amount of micro-vessels within the matrigel [29-30].

5. Concluding Remarks

Current directions in angiogenesis research have become numerous and exciting around the world. This is approximately because of the role of angiogenesis process where it is very fundamental to many physiological conditions. This event plays an important function in wound healing, tissue repair pregnancy as well as exercising. Yet, the abuse of angiogenesis process could lead to pathological condition for example tumor growth. However, angiogenesis is one of the activities that are central to the development and also tissue maintenance. This is why; researchers need to increase their understanding on angiogenesis although this process is still in its early stages. A better understanding on angiogenesis assays could lead to a profound therapeutic outcome of artificial organs development.

Acknowledgment

The authors would like to acknowledge the financial support from Malaysian Ministry of Higher Education (MOHE) grant (vote# 4F124) and Universiti Teknologi Malaysia Tier-1 Research grant (vote # 03H12).

References:

[1] Folkman, J. (1971). Tumor Angiogenesis.Therapeutic Implication. New English Journal of Medicine. 285, 1182-1186.

[2] Kim, K.M. and Evans, G.R.D. (2005). Tissue Engineering : The Future of Stem Cells. *VIII Stem Cells.* 2, 2-21.

[3] Vagaska B, Bacakova L, Filova E, Balik K. (2010). Osteogenic Cells on Bio-Inspired Materials for Bone Tissue Engineering. Physiological Research. 59, 309-322.

[4] Carmeliet, P and Tessier-Lavigne, M.(2005). Common mechanisms of nerve and blood vessel wiring. Nature. 436, 195-200.

[5] Edgar, L.T., Underwood, C. J., Guilkey, J. E., Howing, J. B., Weiss, J. A. (2014). Extracellular matrix density regulates the rate of neovessel growth and branching in sprouting angiogenesis. PLoS One. DOI:10.1371/journal.pone.0085178.

[6] Bauer, S.M., Bauer, R.J., Velazquez, O.C. (2005). Angiogenesis, Vasculogenesis, and Induction of Healing in Chronic Wounds. *Vascular and Endovascular Surgery.*39, 293-306

[7] Auerbach, R., Lewis, R., Shinners, B., Kubai, L., Akhtar, N. (2003). Angiogenesis Assays : A Critical Overview. *Clinical Chemistry.* 49, 32-40.

[8] Vailhe, B., Vittet, D., Feige, J.J. (2001). In vitro Models of Vasculogenesis and Angiogenesis. *Laboratory Investigation.* 81, 439.

[9] Lanza, R.P., Langer, R.S., Vacanti, J. (2007). Principles of tissue engineering. 3rd ed. Boston : Elsevier Academic Press.

[10] Anderson, S.M., Siegman, S.N., Segura, T. (2011). The effect of vascular endothelial growth factor (VEGF) presentation within fibrin matrices on endothelial cell branching. *Biomaterials.* 32, 7432-7443.

[11] Rimann, M. and Graf-Hausner, U. (2012). Synthetic 3D multicellular systems for drug development. *Current Opinion in Biotechnology.* 23, 1-7.

[12] Geckil, H., Xu, F., Zhang, X., Moon, S., Demirci, U. (2010). Engineering hydrogels as extracellular matrix mimics. *Nanomedicine.* 3, 469-484.

[13] Staton, C.A., Stribbling, S.M., Tazzman, S., Hughes, R., Brown, N.J., Lewis, C.E. (2004). Current methods for assaying angiogenesis in vitro and in vivo. *International Journal of Experimental Pathology.* 85, 233-248.

[14] Donovan, D., Brown, N.J., Bishop, E.T., Lewis, C.E. (2001). Comparison of three in vitro human "angiogenesis" assays with capillaries formed in vivo. *Angiogenesis.* 4, 113-121.

[15] Tahergorabi, Z. and Khazaei, M. (2012). A Review on Angiogenesis and Its Assays. *Iranian Journal of Basic Medical Sciences.* 15, 1110-1126.

[16] Phung, M.W., and Dass, C.R. (2006). In -vitro and In –vivo assays for angiogenesis modulating drug discovery and development. *Journal of Pharmacy and Pharmacology.* 58, 153-160.

[17] Li, J., Zhang, Y.P., Kirsner, R.S. (2003). Angiogenesis in Wound Repair : Angiogenic Growth Factors and the Extracellular Matrix. *Microscopy Research and Technique.* 60, 107-114.

[18] Chung, A.S., and Ferrara, N. (2010). The Extracellular Matrix & Angiogenesis. *Pathway.* 11,1-5.

[19] Davis,G.E. and Senger, D.R. (2005). Endothelial Extracellular Matrix: Biosynthesis, Remodelling, and Functions During Vascular Morphogenesis and Neovessel Stabilization. Circulation Research. 97,1093-1107.

[20] Linnes, M.P., Ratner, B.D., Giachelli, C.M. (2007). A fibrinogen-based precision microporous scaffold for tissue engineering. *Biomaterials.* 28, 5298-5306.

[21] Stephanou, A., Meskaoui, G., Vailhe, B., Tracqui, P. (2007). The rigidity in fibrin gels as a contributing factor to the dynamics of *in vitro* vascular cord formation. *Microvascular Research*. 73, 182-190.
[22] Miller, J.C., Pien, H.H., Sahani, D., Sorensen, A.G., Thrall, J.H. (2005). Imaging Angiogenesis: Applications and Potential for Drug Development. *Journal of the National Cancer Institute*. 97, 172-187.
[23] Alessi, P., Ebbinghaus, C., Neri, D. (2004). Molecular targeting of angiogenesis. *Biochimica et Biophysica Acta*. 1654, 39-49.
[24] Benjamin, CJ., Rebecca, SM., Daniel, LC. (2006). GCSF and AMD3100 mobilize monocytes into the blood that stimulate angiogenesis in vivo through a paracrine mechanism and AMD3100 mobilize monocytes into the blood that stimulate angiogenesis in vivo through a paracrine mechanism. *Blood*. 108, 2438- 2445.
[25] Makoto, N., Kimio, S., Yoshihiro, F., Yoshitaka, I., Yutaka, K., Naoto, I., Kazuo, S., & Hiroaki, S. (2007). Important role of erythropoietin receptor to promote VEGF expression and angiogenesis in peripheral ischemia in mice. *Circulation Research*. 100, 662-669.
[26] Christ, PC., Jacquesmorel, CM., Asifamin, M., Matthew, CA., Lisa, HA., & Alisa, KE. (2001). Evidence of IL-18 as a novel angiogenesis mediator. *The Journal of Immunology*. 167, 1644 1653.
[27] Cole, R.W., Liu, F., Herron, B.J. (2010). Imaging of angiogenesis: past, present and future. *Microscopy: Science, Technology, Applications and Education*. pp.885-895.

Literature Review on Dynamic Cellular Manufacturing System

A Nouri Houshyar[1]

Faculty of Engineering, University Putra Malaysia, Malaysia.
A_nh86@yahoo.com

Z. Leman[2]

Faculty of Engineering, University Putra Malaysia, Malaysia.

H. Pakzad Moghadam[3]

Faculty of Engineering, Tehran University, Iran.

M.K.A.M. Ariffin[4]

Faculty of Engineering, University Putra Malaysia, Malaysia.

N. Ismail[5]

Faculty of Engineering, University Putra Malaysia, Malaysia.

H.Iranmanesh[6]

Faculty of Engineering, Tehran University, Iran.

Abstract. In previous decades, manufacturers faced a lot of challenges because of globalization and high competition in markets. These problems arise from shortening product life cycle, rapid variation in demand of products, and also rapid changes in manufcaturing technologies. Nowadays most manufacturing companies expend considerable attention for improving flexibility and responsiveness in order to overcome these kinds of problems and also meet customer's needs. By considering the trend toward the shorter product life cycle, the manufacturing environment is towards manufacturing a wide variety of parts in small batches [1]. One of the major techniques which are applied for improving manufacturing competitiveness is Cellular Manufacturing System (CMS). CMS is type of manufacturing system which tries to combine flexibility of job shop and also productivity of flow shop. In addition, Dynamic cellular manufacturing system which considers different time periods for the manufacturing system becomes an important topic and attracts a lot of attention to itself. Therefore, this paper made attempt to have a brief review on this issue and focused on all published paper on this subject. Although, this topic gains a lot of attention to itself during these years, none of previous researchers focused on reviewing the literature of that which can be helpful and useful for other researchers who intend to do the research on this topic. Therefore, this paper is the first study which has focused and reviewed the literature of dynamic cellular manufacturing system.

1. Introduction

The Cellular manufacturing system is an efficient application of group technology which is able to mix both specifications of flow line and job shop and be responded to the new requirements of the market and it is able to overcome some of the traditional weakness. Cellular manufacturing system can be considered as a successful application of group technology which is able to fulfil today's requirements [2]. Ariafar [1] expressed some of the main benefits for implementing the CMS in comparison with traditional manufacturing system which are as follows:

1. Reduction in inventory
2. Reducing the work in process
3. Decreasing the set up time and throughput time

In cellular manufacturing system, parts are grouped based on their similarities like shape, tolerance and process plan. These groups of parts will be called part families. There are many machine cell in CMS and in each machine cell, there are dissimilar machines which are dedicated to the manufacture of one or more part families. The CMS implementation leads to reducing the throughput time (by reducing the set up time, wait time, and move time), reducing the material handling cost and inventories and also increasing the market response times [3, 4]. Cellular Manufacturing system has two main elements which have been shown in the Figure 1.

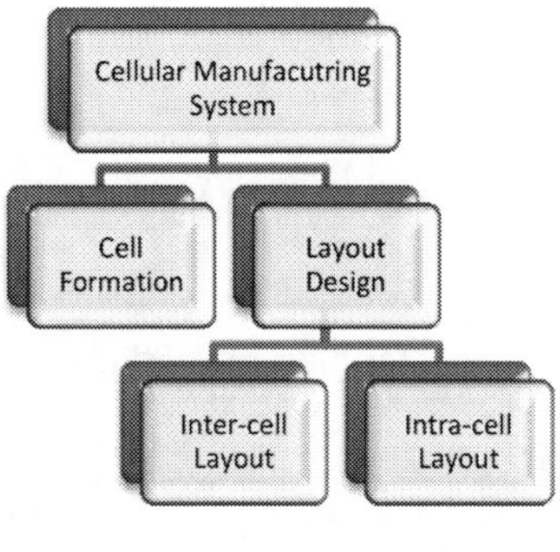

Figure 1, Elements of CMS

The first step in designing CMS is called cell formation (CF) which is process of grouping parts in to the part families and also dedicating part families to machine cell in order to produce. This step in CMS is a crucial stage and gained a lot of attention to itself, indeed it is a well-studied problem during these decades [5, 6, 7, 8]. The next prominent step in designing CMS is layout design. In fact, the aim of this step is, arranging departments, aisles, machines, tools and instruments in the shop floor in the most efficient way. layout design in CMS includes arranging the cells in the shop floor (Inter-cell layout) and also arranging the facilities in the cells (Intra-cell layout). Layout design is an important decision in CMS and has a tremendous impact on manufacturing performance; hence, evaluating the performance of layout is a crucial issue in manufacturing system. In order to evaluate the layout performance, different criteria can be defined for this mean such as: material handling cost, throughput time, lead time and so on. Based on literature, material handling cost is the most important criteria which are applied for evaluating the layout performance. In other words , 20-50 % of the operating expenses have been allocated to material handling cost, and indeed operating expenses can be reduced by 10-30% via an efficient layout design [9]. Therefore, a well-designed layout leads to reduce material handling cost, lead time and throughput time. In contrast, inefficient layout causes to system inefficiency, cost increasing and also accumulated work in process increasing. Although, the importance of layout design is obvious in manufacturing system especially in CMS, this issue absorbed less attention in comparison to cell formation [1, 6, 10, 11, and 12]. On the other hand, another crucial issue in CMS is related to the concept of static or dynamic environment. Indeed, there are two types of manner for cellular manufacturing system which are static and dynamic condition. A static environment is related to the traditional CMS where no changes in the demand are taken into account whereas in dynamic environment, the part demand volume and part mix will change [5]. Most previous articles considered the CMS under the static condition and just formed cells in a single-time period. In contrast, dynamic situation which is more realistic concentrated on a multi-period planning horizon where the demand and product mix will change over the periods; therefore, optimum cell configuration in one period may not be suitable for another period [2]. In addition, due to short life cycle, this leads to changes in demand and product mix over the periods, concentrating on dynamic situation in manufacturing system is more real than static situation. Nowadays most of the researchers focused on dynamic CMS since it is more practical and is more close to reality. Cellular manufacturing system and dynamic environment have been emerging topics in the recent years and they have gained a lot of interest from researcher's side [1].

2. Dynamic cellular manufacturing system

There are two types of manner for cellular manufacturing system which are static and dynamic condition. A static environment is related to the traditional CMS where no changes in the demand are taken into account whereas in dynamic environment, the part demand volume and part mix will change [5]. Most previous articles considered the CMS under the static condition and just formed cells in a single-time period. In contrast, dynamic situation which is more realistic concentrated on a multi-period planning horizon where the demand and product mix will change over the periods; therefore, optimum cell configuration in one period may not be suitable for another period [2]. In addition, due to short life cycle, this leads to changes in demand and product mix over the periods, concentrating on dynamic situation in manufacturing system is more real than static situation. Nowadays, most of the researchers focused on dynamic CMS since it is more practical. For instance, some of the done researches in this area are summarized in this section.

Wicks and Reasor [13] proposed a mixed integer program for solving cell formation problem under dynamic situation in cellular manufacturing system. Their goal were minimizing inter cell material handling cost, reconfiguration cost and capital investment. They applied genetic algorithm for solving their model. In addition, Mungwattana [14] developed two models for CMS under dynamic situation with regards to stochastic production requirement, routing flexibility and cell reconfiguration period by period. Plus, machine allocation problem had been considered in their models for make it more comprehensive. Moreover, Tavakkoli-Moghadam et al., [15] developed cell formation model under the dynamic situation in CMS. In addition, they proposed three meta-heuristic methods including Genetic algorithm, Simulated annealing and Tabu search for solving their problem; then, they compared the performance of their proposed methods. Balakrishnan and Cheng [16] presented a framework for redesigning cells in dynamic cellular manufacturing system. Their framework used a two stages procedure based on generalized machine assignment problem and dynamic programming. Their framework was used for creating a multi-period cellular layout plan for CMS. Defersha and Chen [17] presented a mathematical model for solving cell configuration problem under dynamic situation in CMS. Their model considered alternative routing, lot splitting, operation sequences, machine capacity, work load balancing among cells, operation cost, subcontracting cost, tool consumption cost, set up cost, cell size limits and machine adjacency constraints. Furthermore, Saidi-Mehrabad and Safaei [18] proposed a model for creating cell in CMS under the dynamic situation. Their objectives were, minimizing machine cost, relocation cost and inter-cell movement cost. Neural network was applied for solving their model. Safaei et al., [19] developed a programming model which concentrated on sequence of operations, alternative process plans and machine replication for designing the CMS. They carried out a research to reach the minimum machine fix and variable cost, inter and intra-cell material handling cost and also reconfiguration cost under the dynamic situation. Additionally, Defersha and Chen [20] proposed a mathematical model for solving the dynamic cell formation problem. In their model they considered the accessibility of the machines and tools of the operation. Tavakkoli- Moghadam et al., [21] proposed an integer-linear programming model in order to solve cell formation problem under the dynamic situation in CMS. They applied simulated annealing for solving their problem. They made attempts to reach minimum sum of the inter-cell movement cost and machine cost. In addition, Safaei and Tavakkoli-Moghadam [22] integrated cell formation and production planning problem in a mathematical model under the dynamic situation in CMS. Their research goals were, minimizing inter and intra-cell movement, reconfiguration, subcontracting and inventory holding cost. Bajestani et al., [23] proposed non-linear model for solving cell formation problem and also defining reconfiguration plan in dynamic cellular manufacturing system. Their multi-objective model includes minimization of cell load variation and also minimization of inter-cell material handling cost, reconfiguration cost and machine purchase cost. Kioon et al., [24] presented a mathematical model in order to configure the cell in CMS. They took in to the account different type of production programing such as sequence of operations, dynamic configuration of the cell, multiple process paths, parts outsourcing, maximum number of accessible and limitation of cell size. Moreover, Deljoo et al., [25] solved cell formation problem under dynamic environment in CMS. They applied Genetic Algorithm for solving their model. Mahdavi et al., [26] proposed a model in order to solve the cell formation problem with regards to inventory level, cell reconfiguration and worker assignment to the jobs in dynamic cellular manufacturing system. Their objective were: minimization of holding cost, backorder cost, inter-cell material handling cost, machine cost, reconfiguration cost, salary cost, hiring cost and firing cost. They applied branch and bound for solving their model. Saxena and Jain [27] presented a mixed integer programming model for dynamic CMS in order to solve the cell formation problem. Their model concentrated on machine breakdown, inventory holding, production planning, internal production vs. part outsourcing, inter and intra-cell travel, lot splitting, alternative process routing, operation sequencing, identical machines. Also their model considered machine capacity, tool requirement, work load balancing, machine adjacency constraints, machine procurement and cell reconfiguration. Rafiee et al., [2] proposed a mathematical model which is integration of cell formation and inventory lot sizing problem. They made attempts to minimize the cost of: machine procurement, cell reconfiguration, preventive and corrective repairs, inter and intra-cell material handling, machine operation, part subcontracting, finished and unfinished parts inventory, defective parts replacement. The proposed model was under the dynamic situation with regards to alternative routing, machine capacity limitation, operation sequences, cell size constraints, process deterioration and also the machine break down were taken into account. Moreover, Ghotboddini et al., [28] proposed a mixed integer mathematical model for CMS by considering the dynamic environment. Their model focused on determining the part and machine grouping with labour assignments. Their objectives were, minimizing fixed and variable cost, inter and intra-cell movement, overtime working cost, labour transfer cost, machine purchase cost and also maximizing the human resource utilization rate. Kia et al., [29] proposed an integer non-linear model for designing machine layout under dynamic situation in CMS. They considered different features in their model such as: operation sequence, operation time, alternative process routing, duplicate machine, machine capacity, route selection, production volume of parts, cell reconfiguration, and uncertain part demand and machine capacity. In addition they take into the account the dimensions of machines for making their model more practical. They assumed that all machines have equal-area size. Kia et al., [30] proposed a mixed integer nonlinear programming model for layout design in DCMS. They solved their model via simulated annealing method. Moreover, in their model different elements such as alternative process routing, lot splitting and flexible reconfiguration had been considered. They carried out a research to minimize the cost of: intra and inter-cell material handling, machine relocation, purchasing machines, machine overhead and processing. Majazi Delfard [31] proposed a nonlinear mathematical model for cell formation problem in dynamic environment with regards to number and average length of inter and intra-cell movements in manufacturing system.

3. Conclusion

Cellular manufacturing system is one of the main applications of group technology which gains a lot of attentions to itself during these years. In addition, due to short life cycle which leads to changes in demand and product mix over the periods, concentrating on dynamic situation for manufacturing system is closer to the reality than the static situation; therefore, nowadays most of the researchers focused on dynamic CMS since it is more practical. To put differently, cellular manufacturing system and dynamic environment have been emerging topics in recent years and they have gained a lot of intrest from researchers side; hence, author made attempts to have a brief review on dynamic cellular manufacturing system and focused on all published papers in this topic which can be helpful for other researchers who intend to do the research on this subject.

4. References

[1] Ariafar, S. (2012). Inter-cell and intra-cell facility layout models under different demand environments in cellular manufacturing systems. (PHD thesis), University Putra Malaysia.

[2] Rafiee, K., Rabbani, M., Rafiei, H., & Rahimi-Vahed, A. (2011). A new approach towards integrated cell formation and inventory lot sizing in an unreliable cellular manufacturing system, *J. Applied Mathematical Modelling,* 35(4), 1810-1819.

[3] Wemmerlov, U. and Johnsoon, D.J. (1997). Cellular manufacturing at 46 user plants: implementation experiences and performance improvements. . *Int. J. of Production Resaerch*, 35(1), 29-49.

[4] Wemmerlov, U. and Johnsoon, D.J. (2000). Emperical findings on manufacturing cell design. *Int. J. of Production Research*, 38(3), 481-507.

[5] Papaioannou, G.and Wilson, J.M. (2010). The evoultion of cell formation problem methodologies based on recent studies (1997-2008): Review and directions for future resaerch. *European Journal of Operational Research,* 206(3), 509-521.

[6] Sarker, B.R. and Xu, Y. (1998). Operation sequences-based cell formation methods: A critical survey. *J. Production Planning and Control.,* *9*(8), 771-783.

[7] Yin, Y. and Yasuda, K. (2005). Similarity coefficient methods applied to the cell formation problem: A comparative investigation. *J. Computeres and Industrail Engineering, 48*(3), 471-489.

[8] Yin, Y. and Yasuda, K. (2006). Similarity coeffiecient methods applied to the cell formation problem: A taxonomy and review.. *Int. J. of Production Economics, 101*(2), 329-352.

[9] Tompkins, J. , White, J., & Bozer, Y. (2010). *Facilities Palnning.*

[10] Ahi, A. , Aryanezhad, M.B. , Ashtiani, B. , & Makui, A. (2009). A novel approaach to determine cell formation, intracellular machine layout and cell layout in CMS problem based on TOPSIS method. *J. Computers and Operations Research, 36*(5), 1478-1496.

[11] Sangwan, K.s. and Kodali, R. (2009). FUGEN: A tool for the design of layouts for cellular manufacturing systems. *Int. J. of Services and Operations Mangement, 5*(5), 595-616.

[12] Wang, T.Y., Wu, K.B. and Liu, Y.W. . (2001). A simulated annealing algorithm for facility layout problem under varibale demand in Cellular Manufacuring Systems. *J. Computers in Industry, 46*(2), 181-188.

[13] Wicks, EM., & Reasor, RJ. (1999). Designing cellular manufacturing systems with dynamic part populations. *Inst Indust Eng* 31(1), 11-20.

[14] Mungwattana, A. (2000). Design of cellular manufacturing system for dynamic and uncertain production requirements with presence of routing flexibility. (PhD thesis).

[15] Tavakkoli-Moghaddam, R., Aryanezhad, M. B., Safaei, N., & Azaron, A. (2005). Solving a dynamic cell formation problem using metaheuristics. *J. Applied Mathematics and Computation, 170*(2), 761-780.

[16] Balakrishnan, J., & Cheng, C.H. (2005). Dynamic cellular manufacturing under multiperiod planning horizons. *J. of Manufacturing Technology Management, 16*(5), 516-530.

[17] Defersha, F, & Chen, M. (2006). A comprehensive mathematical model for the design of cellular manufacturing systems. *Int. J. of Production Economics, 103*, 767-783.

[18] Saidi-Mehrabad, M. , & Safaei, N. A. (2007). New model of dynamic cell formation by a neural approach. *Int. J. of Advanced Manufacturing Technology, 33*(9-10), 1001-1009.

[19] Safaei, N., Saidi-Mehrabad, M., & Jabal-Ameli, M. S. (2008). A hybrid simulated annealing for solving an extended model of dynamic cellular manufacturing system. *European Journal of Operational Research, 185*(2), 563-592.

[20] Defersha, F.M., & Chen, M. (2008). A linear programming embedded genetic algorithm for an integrated cell formation and lot sizing considering product quality,. *European Journal of Operation Research 187*, 46–69.

[21] Tavakkoli-Moghaddam, R., Safaei, N., & Sassani, F. (2008). A new solution for a dynamic cell formation problem with alternative routing and machine costs using simulated annealing. *J.of Operational Research Society, 59*(4), 443-454.

[22] Safaei, N. , & Tavakkoli-Moghaddam, R. (2009). Integrated multi-period cell formation and subcontracting production planning in dynamic cellular manufacturing. *Int. J. of Production Economics*.

[23] Bajestani, AM., Rahimi-Vahed, AR., & Khoshkhou, BG. (2009). A multi-objective scatter search for a dynamic cell formation problem. *J. Computer Operation Resaerch, 36*(3), 777-794.

[24] Kioon, S.A., Bulgak, A.A., & Bektas, T. (2009). Integrated cellular manufacturing system design with production planning and dynamic system reconfiguration. *European Journal of Operation Research, 192*, 414-428.

[25] Deljoo, V., Mirzapour Al-e-hashem, S. M. J., Deljoo, F., & Aryanezhad, M. B. (2010). Using genetic algorithm to solve dynamic cell formation problem. *J. Applied Mathematical Modelling, 34*(4), 1078-1092.

[26] Mahdavi, I., Alaei, A., Paydar, MM., & Solimanpour, M. (2010). Designing a mathematical model for dynamic cellular manufacturing systems considering production planning and worker assignment. *J. Computer Math Application, 60*(4), 1014-1025.

[27] Saxena, L., & Jain, P. (2011). Dynamic cellular manufacturing systems design—a comprehensive model. *Int.J. of Advanced Manufacturing Technology 53*(1), 11-34.

[28] Ghotboddini, M. M., Rabbani, M., & Rahimian, H. (2011). A comprehensive dynamic cell formation design: Benders' decomposition approach. *J. Expert Systems with Applications, 38*(3), 2478-2488.

[29] Kia, R., Paydar, M.M., Jondabeh, M.A, Javadian, N., & Nejatbakhsh, Y. (2011). A fuzzy linear programming approach to layout design of dynamic cellular manufacturing systems with route selection and

[30] Kia, R., Baboli, A., Javadian, N., Tavakkoli-Moghaddam, R., Kazemi, M., & Khorrami, J. (2012). Solving a group layout design model of a dynamic cellular manufacturing system with alternative process routings, lot splitting and flexible reconfiguration by simulated annealing. *J. Computers & Operations Research, 39*(11), 2642-2658.

[31] Majazi Dalfard, V. (2013). New mathematical model for problem of dynamic cell formation based on number and average length of intra and intercellular movements. *J. Applied Mathematical Modelling, 37*(4).

Multi-Criteria Knapsack Problem for Disease Selection in an Observation Ward

N Lurkittikul[1] **and O Kittithreerapronchai**[1]

[1] Department of Industrial Engineering, Chulalongkorn University, Bangkok, Thailand 10330

E-mail: nopparuth.l.ty127@gmail.com

Abstract. The aging population and the introduction of Thailand universal healthcare have increased inpatients and outpatients to public hospitals, particularly to a hospital that provides special and comprehensive health services. Many inpatient wards have experienced large influx of inpatients as the hospitals have to admit all patients regardless their conditions. These overcrowding wards cause stress to medical staffs, block access between medical departments, hospital-acquired infections, and ineffective uses of resources. One way to manage such inundated inpatient is to select some patients whose conditions require less clinical attention or whose lengths of stay are predictable and short and, then, place them at an observation ward. This intermediate ward increases turnover of beds and reduces unnecessary paperwork as patients are considered to be outpatients. In this article, we studied inpatient data of a tertiary care hospital in which an observation ward was considered to alleviate the overcrowding problem at Internal Medicine Department. The analysis of data showed that the hospital can balance inpatient flow by managing a group of patients who is admitted because of treatments ordered by its special clinics. Having explored several alternatives, we suggested patient selection criteria and proposed a layout at an observation ward. The hospital should increase medical beds in a new building ward because the current observation ward can handle 27.3% of total short stay patients, while the observation ward is projected to handle 80% of total short stay patients.

1. Introduction

The aging population and the introduction of Thailand universal healthcare have increased numbers of inpatients and outpatients in public hospitals for the last decade. According to Bureau of Policy and Strategy Ministry of Health, the numbers of inpatients and outpatients have approximately doubled from five millions inpatients and 90 million outpatients in 2001 to twelve millions inpatients and 170 million outpatients in 2011, respectively. In particular, public tertiary-care hospitals that operate by the government and provide comprehensive health services have experienced large influx of inpatients as the hospitals are equipped with sufficient resources and have to admit all patients regardless their conditions. These overcrowding wards cause stress to medical staffs, access block between medical departments, hospital-acquired infections, and ineffective uses of resources. All of which undermines the public health quality in terms of prevention and treatment.

Many researchers and health offices have addressed this patient overcrowded problem and proposed recommendations, such as shortening patient's length of stay by streamlining non-medical processes, reducing waiting time by scheduling and preparing key resources, and managing the flow of patients by grouping inpatients by their conditions and/or symptoms. One way to manage such inundated inpatients while accomplishes these three recommendations is to select some patients whose conditions require less clinical attention or whose lengths of stay are predictable and short and, then, place them at a special area, called an observation ward. The observation ward can be viewed a buffer ward between Outpatient Department and Inpatient Department as targeted patients usually stay in the ward during 8 - 48 hours. This intermediate ward increases turnover of beds as it promotes the rapid flow of inpatients and reduces unnecessary paperwork as patients are considered to be outpatients. In some case, an observer ward could reduce the waiting time of patients and the cost of treatments.

The concept of an observation ward – providing specialized services to selected patients so that the patients receive better services in terms of treatment and prevention– is a rational one. In fact, many suggestions and case studies have been reported in the area of Emergency Medicine.

2. Literature review

The relevant researches to this article can be grouped into two categories: observation ward and analytic model in health service.

2.1. Observation ward

An observation ward –also referred to as an observation unit or an observation room– is a health facility located outside inpatient wards that consists of beds and medical equipment designed for providing a short term therapy and/or observing and evaluating symptoms of selected patients [2]. Establishing an observation ward benefits three stakeholders:

- Patient aspect: Recuperating at an observation ward improves the patient's quality of life and satisfaction. Patients who recover in an observation ward usually stay in a hospital shorter than those who stay in a general ward because the patients experience less hospital-acquired infection and receive prompt treatments [2, 5]. Some researchers reported that an observation ward helps reduce the patient mortality [7].
- Medical staff aspect: As patients who recover in an observation ward usually have mild symptoms or minor injuries, the comprehensive investigations of patients, e.g. full triage cross examination, and patient relative interview, are not required [2, 4, 5]. As a result, medical staffs, particularly nurses, spend more time tending patient's needs that, in turn, helps detecting any complications.
- Hospital aspect: A hospital could reduce overall expenditure as patients recover faster with less infection [5, 6].

Nevertheless, an observation ward has drawbacks such as increasing of re-admission rate [6, 7] and demanding additional space and staffs if it is poorly managed. Researchers reported the abuses of an observation ward on difference purposes, such as to prolong length of stay of inpatients, to bill emergency visits, and to handle post-operation patients. Therefore, it is important to establish the clear goals. According to Brillman *et al.* [1], the goals of observation ward are usually intertwined with a medical department that is directly responsible for an observation ward and can be grouped into three types as follows:

- Emergency Department: Majority of observation wards are located in Emergency Department and leaded by emergency physicians to assist patients who require further treatments or evaluations. The main objectives are to improve quality of treatments and to assess conditions of patients. The efficient treatment and effective assessment should reduce admitted patients, waiting times, and cost of treatments.
- Outpatient Department: Some observation ward aims at patients who wait for admission or transferring. This ward is generally operated by non-emergency physicians. Located in

Outpatient Department, its main goal is to manage patients who are medically stable or ready for disposition.

- Inpatient Department: Aimed at cost reduction and alleviated crowded admitted patients, the observation ward located in Inpatient Department targets short stay patients. A group of inpatient physicians and emergency physician are usually in charges of these units to ensure uninterrupted flows of patients.

In some cases, an observation ward could use to observe and to diagnose patients whose causes of disease are unknown or as a temporary ward for treating short stay patients or evaluating the patient's psychosocial needs.

Beside the direct responsible department that usually contributes to the location of an observation ward, the other important parts of establishing the clear goals are the targeted patients and the appropriated time limit. The targeted patients should exhibit clinical conditions easy to observe and treat [1-3] e.g., patient's consciousness, strong vital signs, exact goal of treatment, and mild symptom. An observation ward has to operate within an appropriate time limit to ensure the constant and uninterrupted flow of patients. As a result, many studies suggested different durations in which patients should stay in an observation ward ranging from 8 - 48 hours and depending on the goals. Nevertheless, the previous studies agreed that patients who stay more than 72 hours should be admitted into an inpatient ward [5] because their presence increases the workload and diminishes the effectiveness of an observation ward. The effectiveness of an observation ward depends largely on sufficient resources, such as basic equipment and versatile medical staffs, as patients share similar non-chronic symptoms and an observation ward has to handle patients as a one-stop service.

2.2. Analytic model in health service

A health system is a dedicated and complex system that usually involves the life of human beings. As a result, any initiative faces challenges and securities from stakeholders resulting strong resistance. One way to understand the ramifications of an initiative before the actual implementation is to use an analytic model, such as simulation models and mathematical models.

Simulation models have been widely applied to the health system for different purposes. To narrow the scope of this review, we focus on the simulation models to determine numbers of resources and allocation policy. One of the most competing resources in hospital is an operation room as a surgery may be critical to a patient's life and requires many resources, for example surgeons, anesthesiologists, scrub nurses, and instrument. As a result, the maximum numbers of patients who can be prepped for surgeries was proposed by Ballard *et al* [9] to improve the patient flow while maintain the hospital standards. The simulation leaded to significant numbers of patients who receive surgeries before the original planned. Besides the resources, the workload balancing and bottleneck elimination are important issues in an operation room. Therefore, Zheng *et al* [13] proposed a simulation model to improve process and utilization of key resources, including manpower and facilities. All these studies discover that a simulation model is fundamental for re-engineering and understanding what-if scenario.

Since the simulation model requires enormous data and statistics, one of sensible approaches is a mathematical model embedded with historical data. The inspiration of this approach derives from engineering disciplines, particularly factory and plant design. For example, Caputo *et al* [10] studied the environmental situation at an industrial plant that emits many hazard materials and the decisions by a safety manager to reduce overall hazard released in order to comply with environmental regulations. Therefore, the portfolio of safety measures with budget constraints is modeled as multi-criteria knapsack model. The sensitivity analysis of the model with respected to budgets reveals the useful information for further analytic decision. Another example is the study of foundry industry by

Camargo *et al* [11]. As one of the important Brazilian industries, the key to sustain growth and to maintain competitiveness is the efficient production planning as production sequence and quantities highly affect operation costs. To solve this generalized lot-sizing problem, the two-step heuristic method that combines a genetic algorithm and a knapsack algorithm is proposed and compared with the optimal solution. The result is applicable to the industry as the quality of solution and the computation time are both within reasonable range.

This article aims to illustrate the analysis and procedures necessary for design an observation ward. Having addressed the contribution, the remaining sections are organized as follows. Section 3 overviews the background of the case study hospital along with the analysis of data that lead to the hospital decision to consider an observation ward. Section 4 describes the mathematical model that enables the selection of diseases for an observation ward and the design of an observation room based on available space. Section 5 discusses the conclusion and guidelines to the future research.

3. Case Study Hospital

The case study hospital is the 855-bed tertiary care hospital located in the western region of Thailand. The hospital covers 14 medical specialties for outpatients and 10 medical specialties for inpatients and is capable to handle 600,000 outpatients and 44,000 inpatients annually as shown in figure 1.

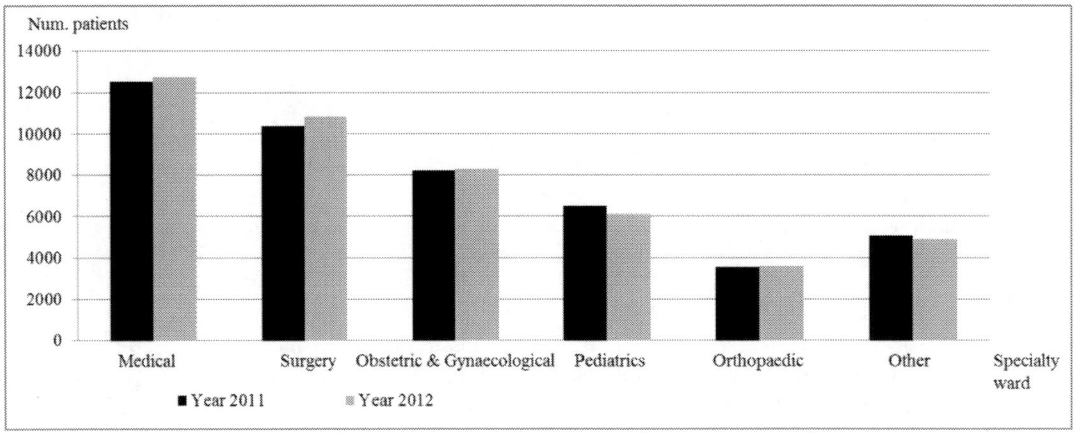

Figure 1. Numbers of patients admitted into specialty wards at the case study hospital in 2011 and 2012.

The figure shows numbers of inpatients in each specialty ward in 2011 and 2012. With exceptions of Pediatrics wards, numbers of patients admitted into major specialty wards increase in 2012. This aligns with the high aging population and low birthrate in the western region. Figure 1 also suggests that 30% of total inpatients are admitted into Internal Medicine Department. This department can be further divided into nine wards as shown in figure 2.

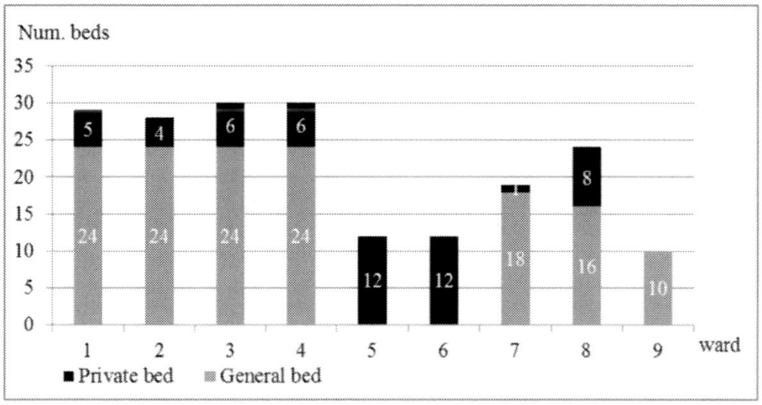

Figure 2. Number of hospital beds at each ward is responsible by Internal Medicine Department.

Figure 2 depicts numbers of registered beds at each ward responsible by Internal Medicine Department which can be classified by expenditure into two types: private bed and general bed. Private beds are dedicated beds located in separated rooms designed to serve patients who prioritize privacy and capable to afford additional expense, whereas general beds are economic beds located in common area and easy to access by nurses. Wards 1, 2, 3 and 4 are main wards of the department as their targeted patients are those who recuperate from incentive care units or have chronic diseases, such as Cancel, Pneumonia, and Leukemia. As the tertiary health facility and the central hospital in the region, the hospital has to admit all patients regardless their conditions, including a large numbers of non-chronic patients. As a result, the numbers of inpatients in these main wards exceed the numbers of registered beds, and the hospital has to use stretchers as temporary medical beds and place non-chronic patients alongside corridors. To investigate this observation, we analyzed the length of stay of inpatient in Internal Medical Wards 1, 2, 3, and 4 in 2011 and 2012, as shown in figure 3.

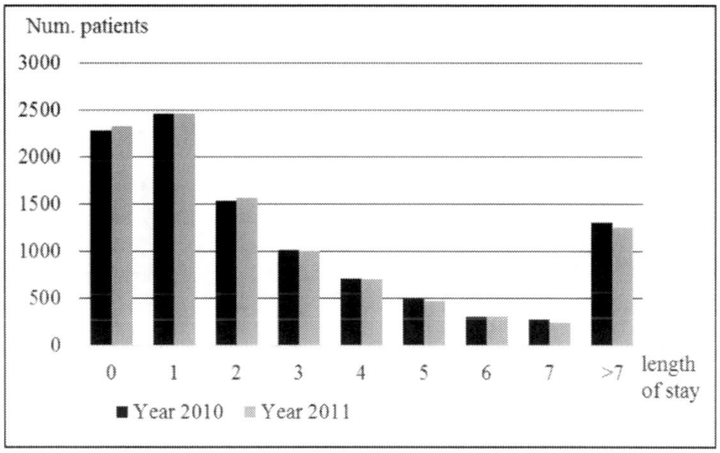

Figure 3. Length of stay of inpatient in Internal Medical Wards 1, 2, 3, and 4 in 2011 and 2012.

Figure 3 confirms our observation. That is, a large portion of inpatients stays in the wards less than two days or 48 hours. In addition, numbers of inpatients who recuperate more than two day gradually decrease. Evidently, these medical wards serve more non-chronic patients than their originally designed, and reducing number of short-stay inpatients should alleviate the overcrowded wards and improve the quality of treatments. However, it is worth to understand arrival patterns of these short-stay inpatients before suggest any solutions, as shown in figure 4.

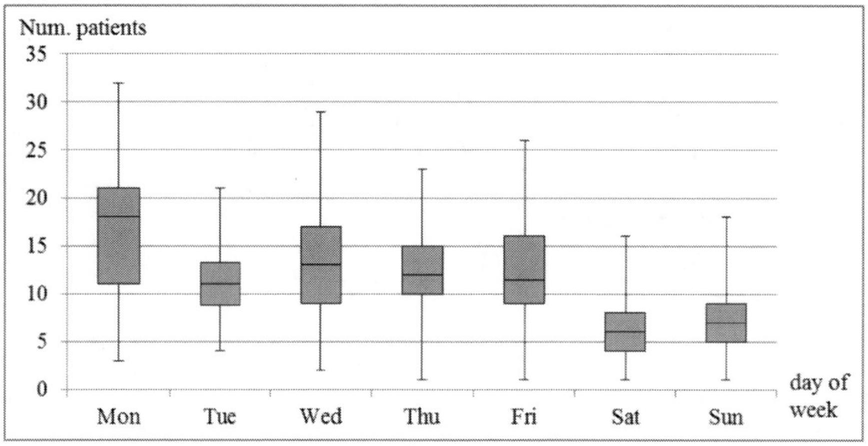

Figure 4. Arrival patterns of inpatients distributions of short stay patient demand on each day of week.

Figure 4 illustrates the arrival patterns and frequencies of the short-stay inpatients on each day of the week. Despite relatively high variations, numbers of the patients who arrive on weekend are less than those who arrive on weekday. The figure shows that the inpatients tend to arrive on Monday or with average of 16.6 and standard deviation of 6.5 patients. The similar pattern can be observed on Wednesday and Friday. The further investigation and the interview with stakeholders reveal that the pattern is influence by special clinics that require specific treatments such as blood transfusion and bone marrow extractions. Moreover, these patients visit the hospital at fixed and predictable time intervals. Having contemplated these evidences and organized series of meetings, the hospital suggests that establishing an observation ward should help to alleviate numbers of short-stay patients in these wards and to streamline medical documents required for these inpatients.

4. Designing an observation ward

After reviewing the purpose and resources required, an observation ward is decided to be located in Emergency Department as shown in figure 5.

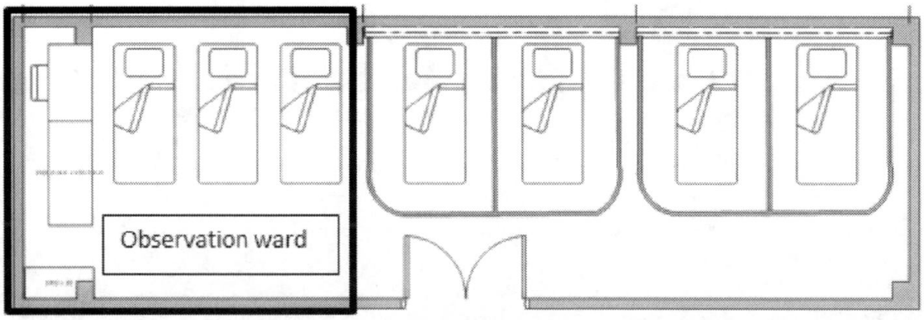

Figure 5. Layout and numbers of medical beds of the proposed observation ward in Emergency Department.

Because of the limited space in Emergency Department, this observation ward contains only three medical beds and serves as a temporary faculty until a new building is constructed and fully functioned. During the meetings, some topics such as availability of nurses, information flows, and supporting activities are raised. Nevertheless, the two biggest concerns are: utilization of medical beds in an observation ward and selection of patients. The answers to these concerns are important and paramount to the success of an observation ward.

Before applying an analytical model to answer these concerns, it is important to understand the nature of length of stay. Unlike the processing time in manufacturing machines, two identical patients who are admitted and diagnosed with the same disease may have different length of stay as illustrated in figure 6.

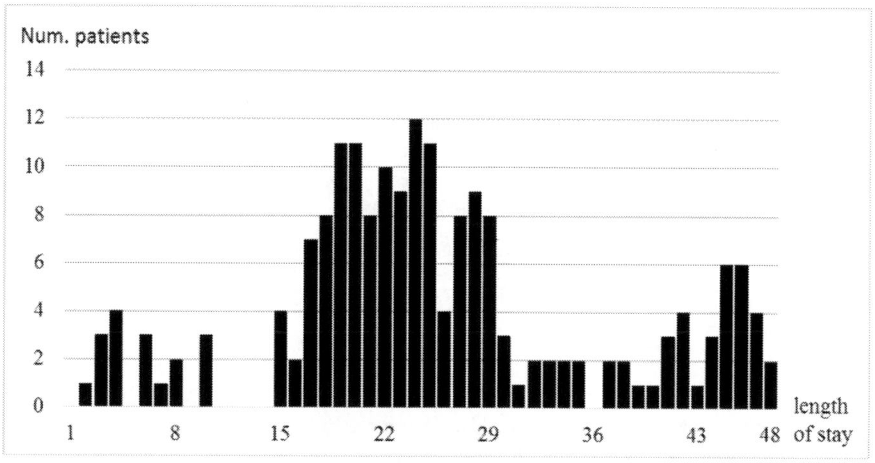

Figure 6. Length of stay of Essential hypertension disease (ICD10:I10).

Figure 6 depicts hours in which patients who are admitted with Essential (primary) hypertension disease spend in the wards. The figure suggests that the length of stay depends on an individual patient. To narrow diseases before using an analytical model, we applied Pareto principle or 80/20 rule to the 2011-2012 medical records. Among 793 diseases that patient spend less than 48 hours in the wards, 114 diseases account for 80% of total short stay patients. These diseases become candidates to an observation ward, and theirs data are embedded in the following mathematical model.

4.1. Mathematical Model

$\mathbb{D}$ = set of diseases

$\mathbb{W}$ = set of day in week

p_d = impact in disease d

l_d = average length of stay of disease d

n_{dw} = average numbers of patient of disease d on day w

H_w = available hours of each medical bed on day w

b = number of medical beds in an observation ward

$$x_d = \begin{cases} 1, & \text{if disease } d \text{ is selected in the observation ward} \\ 0, & \text{otherwise.} \end{cases}$$

$$\max \ z = \sum_{d \in \mathbb{D}} p_d \cdot x_d \tag{1}$$

$$\text{s.t.} \ \sum_{d \in \mathbb{D}} l_d \cdot n_{dw} \cdot x_d \leq b \cdot H_w \qquad \forall w \in \mathbb{W} \tag{2}$$

$$x_d \in \{0,1\} \qquad \forall d \in \mathbb{D} \tag{3}$$

Expression 1 is the objective function that maximizes the total impacts of the selected diseases. The hospital is interested in impacts as many criteria such as numbers of patent in each disease, possibility to develop serious symptom, nature of treatment, available equipment and skilled of staff. These criteria are different from one hospital to others and require serious discussion among physicals and medical staffs. The constraints of this Integer Programming are listed in Expressions 2 and 3. Expression 2 ensures that the average duration required by patients who are served by an observation ward on each day of the week does not excess the medical-bed capacities. This expression reflects the fact that medical beds are allocated directly to each department and cannot be easily borrowed or transferred. Expression 3 assures that the decision variable or the selection of disease is binary as an initial triage unable to predict the length of stay of each patient. Mathematically, Expressions 1 – 3 can be viewed as the multi-criteria knapsack problem – a generalization of classical knapsack problem in which a person must select a set of items to maximize total benefits while satisfy the limitations of all resources. Using the analogy of the multi-criteria knapsack problem, parameters p_d, $l_d \cdot n_{dw}$, and $b \cdot H_w$ can be viewed as the benefit of each item, resources required of each item, and total limited resources, respectively.

4.2. Numerical Experience

We embedded the data of the case study hospital into our model and solved it with Microsoft Excel/ Solver [12]. Since the available hours of each medical bed is a difficult parameter to be estimated, we varied the parameter from 14 hours to 24 hours with two-hour interval, i.e., $\{14,16,18,20,22,24\}$ and reported the result and numerical experience as shown in table 1.

Table 1. Numbers of disease and percentage of served inpatient with different available hours of each medical bed (H_w).

H_w	numbers of diseases ($\sum_{d \in \mathbb{D}} x_d$)	% of served patients [a]	Daily average utilization (%)	Daily standard deviation of utilization (%)
24	42	27.3	93.1	10.4
22	39	25.1	93.4	9.1
20	36	23.3	93.8	9.1
18	29	21.2	93.9	8.9
16	31	19.2	93.5	10.0
14	28	17.1	94.6	8.0

[a] percentage form 80% of total short stay patient (targeted patient)

As the available hours of each medical bed decrease, table 1 suggests that number of patients who are served by an observation ward gradually decreases. Decreasing H_w results in changing the set of diseases. Interestingly, the number of diseases at $H_w = 16$ is 31 diseases which is larger than $H_w = 18$ or 29 diseases because many diseases are introduced into an observation ward at $H_w = 16$. In particular, the model selects diseases to maintain high utilization of an observation ward. It is worth to mention that the model fails to reach 100% utilization as average numbers of patients on Saturday and Sunday are significant less than that on weekday, resulting to 93% utilization of medial beds.

4.3. Greenfield Design

Because of the limited space, the proposed observation ward can handle 17.1-27.3% of potential inpatients. This raises a question: how many medical beds does the hospital require if there is no spatial constraint? As a result, we modified Expression 2 by assuming $x_d = 1 \ \forall d \in \mathbb{D}$ to determine the number of medical beds required with Expression 4.

$$b = \max_{w \in \mathbb{W}} \left\lceil \frac{\sum_{d \in \mathbb{D}} l_d \cdot n_{dw}}{H_w} \right\rceil \tag{4}$$

Number of beds in an observation ward (b) is the maximum round-up value of total hours all short-stay patients spend in an observation ward each day over available hours of bed. The result of this calculation, average utilization, and standard deviation are shown in table 2.

Table 2. Number of beds in an observation ward with different available hours of each medical bed (H_w).

H_w	number of beds (b)	% of served patients [a]	Daily average utilization (%)	Daily standard deviation of utilization (%)
24	17	100	69.4	18.4
22	18	100	71.5	18.9
20	20	100	70.8	18.7
18	22	100	71.5	18.9
16	25	100	70.8	18.7
14	28	100	72.3	19.1

[a] percentage form 80% of total short stay patients

As the available hours decrease, table 2 shows that the numbers of beds gradually increase. The result reveals that the utilization of beds depends on number of patients on Monday. This reflects the fact that majority of patients show up on Monday resulting to the under-utilization in other days. If numbers of inpatients were similar to other days, the observation ward would require only 15 beds instead of 20 beds in case of $H_w = 20$ or 25% less capacity. This observation leads to the discussion on patient flows. The imbalance of patient flows affects to utilization of beds. The average of utilization in Greenfield Design obviously lower than Numerical Experience because the model in Numerical Experience attempt to maximum impact and it affects high utilization simultaneously.

5. Conclusion and future works

We analyzed the data of a Thailand tertiary care hospital and found that the hospital has faced the overcrowded inpatient in the major wards responsible by Internal Medicine Department as the inpatients spends less than 48 hours in the wards. The additional investigation revealed that some of these short-stay patients are caused by special treatments. As a result, the hospital considers an observation ward to alleviate this problem and to streamline procedures. Furthermore, the meeting between Internal Medicine Department and Emergency Department concluded a temporary site of an observation ward. Because of a limited space, we proposed a mathematical model to select the set of diseases and determined numbers of medical beds if there is no spatial constraint. The model suggests that three medical beds available at Emergency Department can ideally handle approximately 27.3% of targeted patients that consists of 42 diseases. Depending on the medical bed turnover, 17-28 medical beds are recommended if the hospital plans to setup a new observation ward in a new building. The result also implies that Internal Medicine Department should review special clinics schedule to balance the patient flows and to reduce a number of beds required in an observation ward.

The proposed mathematical model serves as a primary design tool as the deterministic arrival time and recuperated time are assumed. The further studies and comprehensive design using a discrete event simulation that accounts for the stochastic events are required. In addition, the design of an observation ward should incorporate with the procedure and information flow to ensure smooth and successful implementation.

Acknowledgement
We are forever indebted to Thai Health Promotion Foundation for financial supports and personal connection to the case study hospital. The authors would like to thank Thira Woratanara, M.D.,Wut Dulyachai, M.D., Supachai Paiboonpol, M.D., and the staffs at the case study hospital for their valuable time, useful comments, and various supports throughout this project.

References
[1] Brillman J *et al* 1995 Management of observation units *Annals of Emergency Medicine* **25** 823-830.
[2] Brand C et al 2004 Short Stay and Observation Units, Medical Assessment and Planning Units and Emergency Medical Units. *Clinical Epidemiology and Health Service Evaluation Unit* (Australia: Melbourne Health), Retrived from *http://www.mh.org.au/.*
[3] Dallos V and Mouzas GL 1981 An evaluation of the functions of the short-stay observationward in the accident and emergency department *British Medical Journal* **282** 37.
[4] Jones A, O'Driscoll K and Luke LC 1995 Head injuries and the observation ward *Journal of Accident and Emergency Medicine* **12** 160.
[5] Cooke MW, Higgins J and Kidd P 2003 Use of emergency observation and assessment wards: a systematic literature review *Emerg Med J* **20** 138-142.
[6] Daly S, Campbell DA and Cameron PA 2003 Short-stay units and observation medicine: a systematic review *The Medical Journal of Australia (MJA)* **178** 559-563.
[7] Juan A, Salazar A, Alvarez A, Perez JR, Garcia L and Corbella X 2006 Effectiveness and safety of an emergency department shortstay unit as an alternative to standard inpatient hospitalization *Emerg Med J* **23** 833-837.
[8] Ballard SM and Kuhl ME 2006 *Winter Simulation Conference (WSC)* .
[9] Zheng Q, Shen J, Liu Z, Fang K and Xiang W 2011 *International Conference on Industrial Engineering and Engineering Management (IE&EM).*
[10] Caputo AC, Pelagagge PM and Salini P 2013 A multicriteria knapsack approach to economic optimization of industrial safety measures *Safety Science* **51** 534-360
[11] Camargo VCB, Mattiolli L and Toledo F 2012 A knapsack problem as a tool to solve the production planning problem in small foundries *Computers and Operations Research* **39** 86-92
[12] Fylstra D, Lasdon L, Watson J and Waren A 1998 Design and use of the microsoft excel solver *Interfaces* **28** 29-55.

PH Tester Gauge Repeatability and Reproducibility Study for WO_3 Nanostructure Hydrothermal Growth Process

Amirul Abd Rashid [1,2,a], Nor Hayati Saad [2,b], Daniel Bien Chia Sheng[1,c] and Lee Wai Yee[1,d]

[1]MIMOS Berhad, Technology Park Malaysia, Kuala Lumpur, 57000, Malaysia
[2]Micronanoelectro Mechanical System Laboratory (MINEMS), Faculty of Mechanical Engineering, Universiti Teknologi MARA, 40450 Shah Alam, Selangor, Malaysia

[a]amirul2550@ salam.uitm.edu.my, [b]norhayatisaad@salam.uitm.edu.my, [c]daniel.bien@mimos.my, [d]wy.lee@mimos.my

Abstract. PH value is one of the important variables for tungsten trioxide (WO_3) nanostructure hydrothermal synthesis process. The morphology of the synthesized nanostructure can be properly controlled by measuring and controlling the pH value of the solution used in this facile synthesis route. Therefore, it is very crucial to ensure the gauge used for pH measurement is reliable in order to achieve the expected result. In this study, gauge repeatability and reproducibility (GR&R) method was used to assess the repeatability and reproducibility of the pH tester. Based on ANOVA method, the design of experimental metrics as well as the result of the experiment was analyzed using Minitab software. It was found that the initial GR&R value for the tester was at 17.55 % which considered as acceptable. To further improve the GR&R level, a new pH measuring procedure was introduced. With the new procedure, the GR&R value was able to be reduced to 2.05%, which means the tester is statistically very ideal to measure the pH of the solution prepared for WO_3 hydrothermal synthesis process.

1. Introduction

Tungsten trioxide (WO_3) nanostructure is one of the metal oxide nanostructures that can be used as a functional element in various gas sensor applications [1] to detect gases such as ethylene, hydrogen, nitrogen oxide and others. WO3 is a n-type indirect band gap semiconductor and possesses a band gap energy range from 2.4 – 2.8EV that results in a higher utilization of the solar spectrum [2]. Nanostructure of WO3 can certainly enhance the performance of functional material and provide unique properties that are absent in its bulk form, hence it becomes rapidly expands in the field of materials chemistry [3].

There are few methods to grow WO3 nanostructure includes sol gel, chemical vapor deposition (CVD) and hydrothermal process [4]. Among them, the hydrothermal method received heavy interest due to fact that the preparation and equipment used for this process is relatively simple and less complicated. Hydrothermal is defined as the method to produce different chemical compounds and materials at temperatures above 100°C and pressures above 1 ATM by using physical and chemical processes flowing in aqueous solutions [5]. Typically, hydrothermal treatment begins with the preparation of a precursor solution [6]. The pH precursor is then added with appropriate amount acid or alkaline agent until the desired value is achieved. Depends on final nanostructure required, additional additive such as capping reagents or directive agents was mixed with the prepared solution and heated in a dedicated reactor to certain temperature ranges from 120 – 300 °C for a desired period of time. This allows the nucleation and growth of the crystallites took place.

Previous studies have established a few treatment parameters such as precursor concentration, pH, temperature, and growth duration significantly influence the morphological properties of the nanostructure [7]. Out of those variables, pH alone was discovered as a very important parameter which dictates the shape of the WO3 nanostructures, leading to a modification of the morphology from rod-like to prism-like and flower-like structures as illustrated in Figure 1.

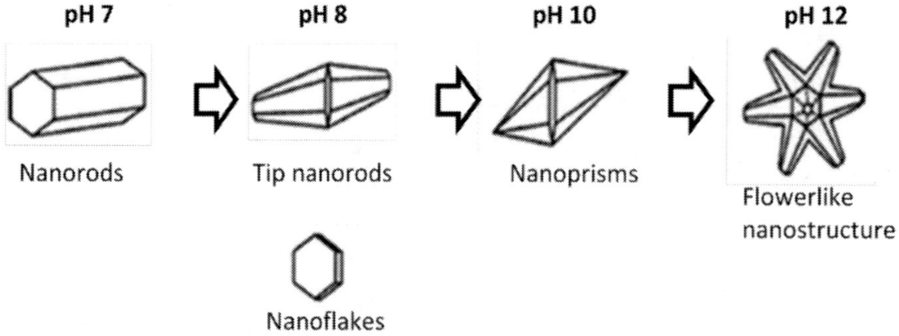

Figure 1. Illustration of different WO_3 morphology synthesis using hydrothermal with different pH value [8].

Therefore, it is very important that the measurement tools used to measure the pH value is accurate and reliable. One of the techniques largely used to access the measurement tool capability is gauge repeatability and reproducibility (GR&R) method. GR&R is a statistical tool that measures the amount of variation in the measurement system arising from the measurement device and the people taking the measurement [9]. Using this method, we can determine the uncertainty of the measurement systems before we can compare, control or optimize the manufacturing processes. Based on analysis of variance (ANOVA) method, the GR&R value of less than 10% is considered ideal while value of 10 - 20% is acceptable but there should be some plan to review the system for improvement. The GR&R value more than 30% indicates that the measurement system under the study is unacceptable which require either more accurate tool to be used or improvement of the measurement procedure is required.

2. Methodology

GR&R study on the PICCOLO pH tester (Hanna Instruments, USA) was conducted according to the process flow illustrated in Figure 2. Using MINITAB software, metrics for three operators measuring different buffer solution was generated. The 10 sample measurement was repeated by the operators in random order to minimize the bias in the analysis. To determine the GR&R value, analysis of variance (ANOVA) method which is also available in MINITAB software was used to analyze the data obtained from the study.

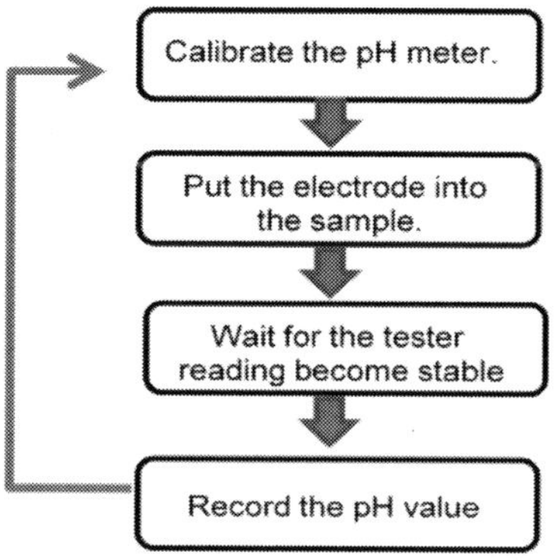

(a) (b)

Figure 2. Positioning of the tester to measure pH value (a) and the repeated process flow to conduct the pH measurement (b).

3. Result & Discussion

ANOVA analysis was done on the collected pH data from the experiment and the distribution of the variation component as shown in Figure 3 (a). It can be seen that the part-to part is the majority that contributes to the variation due to different pH value used as part of the experiment.

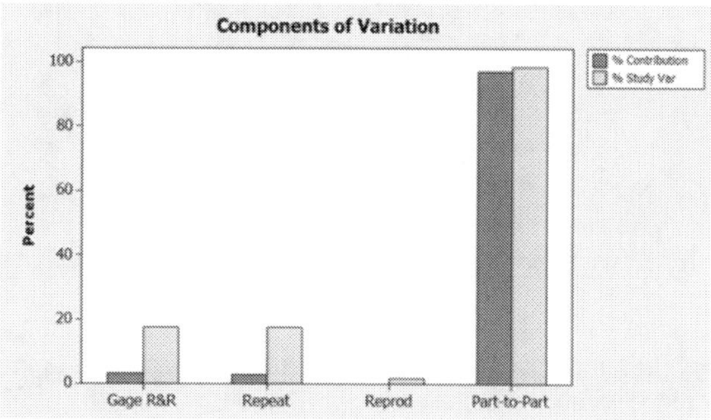

Figure 3. Histogram of variations components

Table 1 shows the ANOVA analysis, which indicates the total GR&R value of 17.55 %. Even though this value is considered acceptable because it is well below the 30 % range, but it is worth to revise the pH testing process to further improve the reliability and consistency of the measurement process.

Table 1. ANOVA analysis results of GR&R study

	StdDev (SD)	Study Var (6 * SD)	% Study Var (% SV)
Total Gage R&R	0.41445	2.4867	17.55
Repeatability	0.41235	2.4741	17.46
Reproducibility	0.04175	0.2505	1.77
Operator	0.04175	0.2505	1.77
Part-to-part	2.32437	13.9462	98.45
Total Variation	2.36103	14.1662	100.00

Base on observations and further discussion, the same GR&R study was carry out again but this time, two simple steps was included as illustrated in Figure 4.

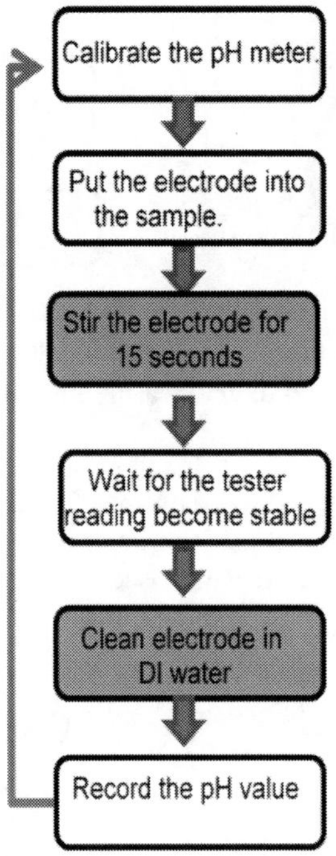

Figure 4. Additional two process steps (highlighted in blue) included in the process flow

The new ANOVA analysis in Table 2 confirms that the new GR&R value has been further reduced to 2.5%. With such low percentage of variation, the pH tester used and in the measurement process is statistically ideal to produce reliable and consistence pH value.

Table 2. ANOVA analysis results with the new additional measurement steps.

	StdDev (SD)	**Study Var (6 * SD)**	**% Study Var (% SV)**
Total Gage R&R	0.05413	0.3248	2.05
Repeatability	0.05280	0.3168	2.00
Reproducibility	0.01192	0.0715	0.45
Operator	0.01192	0.0715	0.45
Part-to-part	2.64508	15.8705	99.98
Total Variation	2.64564	15.8738	100.00

4. Conclusion

The GR&R of the pH tester and measurement procedure was successfully assessed. Based on ANOVA analysis, the initial pH tester GR&R was found to be 17.55 % which is statistically in acceptable condition. By introducing new pH measurement process step, the GR&R was able to further reduce to 2.5 %. The result proved that the tester and method is fit to be used and provides a higher certainty that the pH value measured is accurate. This will contribute to minimize the overall process variation of the synthesized nanostructure produced via hydrothermal method.

Acknowledgement
The authors wish to thank MIMOS Berhad for facility accessibility, Universiti Teknologi MARA (UiTM) and the Ministry of Education Malaysia (MOE) for the financial support under the Exploratory Research Grant Scheme (File No: 600-RMI/ERGS 5/3 (18/2013))

References

[1] Siciliano, T., et al. "WO^3 gas sensors prepared by thermal oxidization of tungsten." Sensors and Actuators B: Chemical 133.1 (2008): 321-326.
[2] Su, Jinzhan, et al. "Vertically aligned WO3 nanowire arrays grown directly on transparent conducting oxide coated glass: synthesis and photoelectrochemical properties." Nano letters 11.1 (2010): 203-208.
[3] Ng, Charlene Jin Wei. Synthesis Of Tungsten Oxide For Solar Energy Conversion And Water Splitting Applications. Diss. The University of New South Wales, 2012.
[4] Zhang, Jun, et al. "Hydrothermally synthesized WO3 nanowire arrays with highly improved electrochromic performance." Journal of Materials Chemistry 21.14 (2011): 5492-5498.
[5] Adschiri, Tadafumi, Yukiya Hakuta, and Kunio Arai. "Hydrothermal synthesis of metal oxide fine particles at supercritical conditions." Industrial & engineering chemistry research 39.12 (2000): 4901-4907.
[6] Rashid, Amirul Abd, et al. "Preliminary Study of WO3 Nanostructures Produced via Facile Hydrothermal Synthesis Process for CO2 Sensing."Applied Mechanics and Materials 431 (2013): 37-41.
[7] G. Amin, M. H. Asif, A. Zainelabdin, S. Zaman, O. Nur, and M. Willander (2011), Influence of pH, Precursor Concentration, Growth Time, and Temperature on the Morphology of WO3 Nanostructures Grown by the Hydrothermal Method.
[8] D. Vernardou, G. Kenanakis, S. Couris, E. Koudoumas, E. Kymakis, N. Katsarakis (2007), pH effect on the morphology of WO3 nanostructures grown with aqueous chemical growth.
[9] Vassilakis, E., and G. Besseris. "An application of TQM tools at a maintenance division of a large aerospace company." Journal of Quality in Maintenance Engineering 15.1 (2009): 31-46.

On the Eulerian recurrent lengths of complete bipartite graphs and complete graphs

Shuji Jimbo

Graduate School of Natural Science and Technology, Okayama University, 3-1-1 Tsushima-naka, Kita-ku, Okayama-shi, Okayama 700-8530, JAPAN

E-mail: `jimbo-s@okayama-u.ac.jp`

Abstract. An Eulerian circuit of a graph is a circuit that contains all of the edges of the graph. A graph that has an Eulerian circuit is called an Eulerian graph. The Eulerian recurrent length of an Eulerian graph G is the maximum of the length of a shortest subcycle of an Eulerian circuit of G. In other words, if every Eulerian circuit of an Eulerian graph G has a subcycle of length less than or equal to l, and there is an Eulerian circuit of G that has no subcycle of length less than l, then the Eulerian recurrent length of G is l. The Eulerian recurrent length of graph G is abbreviated to the ERL of G, and denoted by ERL(G). In this paper, the ERL's of complete bipartite graphs are given. Let m and n be positive even integers with $m \geq n$. It is shown that $\mathrm{ERL}(K_{m,n}) = 2n - 4$ if $n = m \geq 4$, and $\mathrm{ERL}(K_{m,n}) = 2n$ otherwise. Furthermore, upper and lower bounds on the ERL's of complete graphs are given. It is shown that $n - 4 \leq \mathrm{ERL}(K_n) \leq n - 2$ holds for every odd integer n greater than or equal to 7.

1. Introduction

An Eulerian circuit of a graph is a circuit that contains all of the edges of the graph. A graph that has an Eulerian circuit is called an Eulerian graph. Finding an Eulerian circuit of a graph, that is to say drawing the graph with a single stroke of the brush, is a fundamental problem that was studied by Euler in the dawn of graph theory. It is known that a connected graph is Eulerian if and only if every degree of a vertex of the graph is even, and there is a linear time algorithm to find an Eulerian circuit of an Eulerian graph. Algorithms to find an Eulerian circuit are utilized for reconstructing original long base sequences from short fragments of DNA in the field of bioinformatics[1].

We investigate a problem finding an Eulerian circuit such that the length of a shortest subcycle of the Eulerian circuit is as long as possible. We call the problem the Eulerian recurrent length problem (ERLP). The Eulerian recurrent length of an Eulerian graph G is the maximum of the length of a shortest subcycle of an Eulerian circuit of G. In other words, if every Eulerian circuit of an Eulerian graph G has a subcycle of length less than or equal to l, and there is an Eulerian circuit of G that has no subcycle of length less than l, then the Eulerian recurrent length of G is l. The Eulerian recurrent length of graph G is abbreviated to the ERL of G, and denoted by ERL(G). For example, the ERL of the graph in figure 1 that consists of $3n$ vertices is n. We hope that finding the ERL of a graph is useful for solving some optimization problem.

It has been proved that there is no approximation algorithm with a constant approximation ratio for the ERLP[2]. In this paper, the ERL's of complete bipartite graphs are given. Let m and n be integers with $0 < n \leq m$. It is shown that $\mathrm{ERL}(K_{m,n}) = 4n - 4$ if $n = m$, and

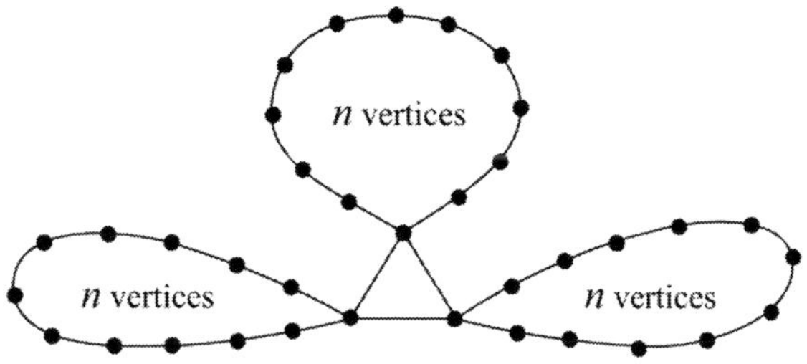

Figure 1. A graph whose ERL equals n.

$\mathrm{ERL}(K_{m,n}) = 4n$ otherwise. Those results are included in our articles[3, 4] written in Japanese. Furthermore, an upper and lower bound on the ERL's of complete graphs are given. It is shown that $n - 4 \le \mathrm{ERL}(K_n) \le n - 2$ holds for every odd integer n greater than or equal to 7. The lower bound slightly improves the previous one[3]. As shown above, the ERL's of complete bipartite graphs and complete graphs are close to trivial upper bounds.

In the next section, we shall define several notions necessary for the arguments that follow. In section 3, we shall give the ERL's of complete bipartite graphs. In section 4, we shall give an upper and lower bound on the ERL's of complete graphs. In the last section, we shall give conclusions and remarks about a further challenge to determine the ERL's of complete graphs.

2. Preliminaries

A walk is an alternating sequence of vertices and edges $v_0 \to e_1 \to v_1 \to \cdots \to v_{k-1} \to e_k \to v_k$, beginning and ending with a vertex, such that, for each $i \in \{1, 2, \ldots, k\}$, v_{i-1} and v_i are both end vertices of e_i, and v_{i-1} is different from v_i if e_i is not a loop. Every graph that appears in this paper is a simple undirected graph. We may, therefore, express a walk W with only its vertices as $v_0 \to v_1 \to \cdots \to v_m$, where v_0 is the initial vertex and v_m the terminal vertex. The walk W is said to be a v_0-v_m walk, or a walk from v_0 to v_m. A walk is said to be closed if the initial and final vertex are identical. If a walk is closed, then the walk is expressed as $W = v_0 \to v_1 \to \cdots \to v_m \to v_0$. In this case, the walk W can be written as $v_i \to v_{i+1} \to \cdots \to v_m \to v_0 \to \cdots v_i$ for each $i = 0, 1, \ldots, m$, since each vertex in the walk can be regarded as the initial and end vertex. The length of a walk is the number of edges in the walk, even if the walk is closed.

A trail is a walk such that all its edges are distinct. A circuit is a closed trail. A path is a trail such that all its vertices are distinct except that its initial and final vertex are identical. A cycle is a closed path that has at least one edge. Let G be a graph, and W_1 and W_2 walks in G. If W_1 is a subsequence of W_2, then W_1 is said to be a subwalk of W_2. If C is a cycle and a subwalk of W, then C is said to be a subcycle of W. The terms, subtrail, subcircuit, and subpath, are also defined in the same manner. We may regard a path of a graph as the subgraph induced from the vertices in the path.

An Eulerian circuit of a graph G is a circuit of G that contains all of the edges of G. A graph is Eulerian if it has an Eulerian circuit. It is a well-known fact that a graph is Eulerian if and only if the degree of each vertex of the graph is even.

3. The Eulerian recurrent length of complete bipartite graphs

The ERL's of complete bipartite graphs are given in the following theorem.

Theorem 1 *Let m and n be positive even integers with $m \ge n$. If $m = n \ge 4$, then $\mathrm{ERL}(K_{m,n}) = 2n - 4$ holds. Otherwise, $\mathrm{ERL}(K_{m,n}) = 2n$ holds.*

Proof. It is clear that $\mathrm{ERL}(K_{2,2}) = 4$ holds. The proof immediately follows from this fact and two lemmas below. □

The following lemma and the proof are both presented in [3].

Lemma 1 *Let n be an even integer greater than or equal to 4. Then,* $\mathrm{ERL}(K_{n,n}) = 2n - 4$ *holds.*

Proof. First, we construct an Eulerian circuit T of $K_{n,n}$ such that the length of a shortest subcycle of T is at least $2n-4$. Without loss of generality, we may assume that the vertex-set of $K_{n,n}$ is $\{0, 1, \ldots, 2n-1\}$ and that all of the edges of $K_{n,n}$ join a vertex in $U = \{0, 2, 4, \ldots, 2n-2\}$ and a vertex in $V = \{1, 3, 5, \ldots, 2n-1\}$. For each even integer $k = 0, 2, \ldots, n-2$, let H_k denote the Hamilton path of $K_{n,n}$ defined as

$$\begin{aligned} H_k &= 0 \to (2k+1) \bmod 2n \to 2 \to (2k+3) \bmod 2n \to \cdots \to 2n-2 \\ &\quad \to (2k+2n-1) \bmod 2n . \end{aligned}$$

The edge-set of $K_{n,n}$ is partitioned into $n/2$ Hamilton cycles C_0, C_2, ..., and C_{n-2}, where each C_k is expressed in the form of $H_k \to 0$. It follows easily from the definition of H_k's that the Hamilton cycles C_0, C_2, ..., and C_{n-2} are edge-disjoint each other. Therefore, the closed walk

$$T = H_0 \to H_2 \to H_4 \to \cdots \to H_{n-2} \to 0$$

is an Eulerian circuit.

Since T is a concatenation of Hamilton cycles, every shortest subcycle of T must be contained in a trail $H_k \to H_{(k+2) \bmod n}$ for some $k \in \{0, 2, \ldots, n-2\}$. We assume that k is an even integer with $0 \le k \le n-2$, and shall show that the length of any subcircuit τ of $H_k \to H_{(k+2) \bmod n}$ is not less than $2n-4$. It follows easily from the definition of H_k's that if i is an even integer with $0 \le i \le n-1$, then the length of the subcircuit of $H_k \to H_{(k+2) \bmod n}$ from the vertex i in H_k to the vertex i in $H_{(k+2) \bmod n}$ is equal to $2n$. We may, therefore, assume that there is an odd integer j such that τ is the subcircuit from j in H_k to j in $H_{(k+2) \bmod n}$.

Let x and y denote the integers defined by

$$H_k = 0 \to \cdots \to x \to j \to \cdots \to (2k+2n-1) \bmod 2n$$

and

$$H_{(k+2) \bmod n} = 0 \to \cdots \to y \to j \to \cdots \to (2k+2n+3) \bmod 2n .$$

Then, it follows immediately from the definition of H_k's that $x = (j - 2k - 1) \bmod 2n$ and $y = (j - 2k - 5) \bmod 2n$. Therefore, the length of the latter part of H_k from j is $l = 2n - ((j - 2k - 1) \bmod 2n) - 2$, and the length of the former part of $H_{(k+2) \bmod n}$ to j is $r = ((j - 2k - 5) \bmod 2n) + 1$. It follows from $n \ge 4$ that if $(j-2k-1) \bmod 2n > (j-2k-5) \bmod 2n$ then $(j - 2k - 1) \bmod 2n = ((j - 2k - 5) \bmod 2n) + 4$. We, therefore, have $l + r + 1 \ge 2n - 4$. Thus, it is concluded that the length of a shortest subcycle of T is not less than $2n - 4$.

Next, we assume that there is an Eulerian circuit T of $K_{n,n}$ that has no subcycle of length less than or equal to $2n - 4$, and shall derive a contradiction. Since the order of $K_{n,n}$ is $2n$, T has a subcycle of length at most $2n$. Suppose that T has a subcycle

$$S = s_1 \to s_2 \to \cdots \to s_{2n} \to s_1$$

of length $2n$. Since any bipartite graph have no cycles of odd length, $E = \{s_2, s_4, \ldots, s_{2n}\}$ and $O = \{s_1, s_3, \ldots, s_{2n-1}\}$ are the vertex classes of $K_{n,n}$, in other words, either $E = U$ and $O = V$ or $E = V$ and $O = U$ holds. Let x_1 and x_2 denote the vertices defined by

$$T = \cdots \to S \to x_1 \to x_2 \to \cdots .$$

It follows from the definition that $x_1 \in E$ and $x_2 \in O$. Any subtrail of T must satisfy the following two conditions.

Table 1. Violations of the conditions when $x_1 \neq s_4$

z	Walk $S \to z$ violates …
s_2, s_{2n}	Condition 1
$s_6, s_8, \ldots, s_{2n-2}$	Condition 2

Table 2. Violations of the conditions when $x_1 = s_4$

z	Walk $S \to s_4 \to z$ violates …
s_1, s_3, s_5	Condition 1
$s_7, s_9, \ldots, s_{2n-1}$	Condition 2

Table 3. Violations of the conditions when $x_1 \neq s_{2n}$

z	Walk $S \to z$ violates …
s_2, s_{2n-2}	Condition 1
$s_4, s_6, \ldots, s_{2n-4}$	Condition 2

Condition 1 No edge appears more than two times in the subtrail.
Condition 2 There is no subcycle of the subtrail of length less than or equal to $2n - 4$.

From table 1, we have $x_1 = s_4$. Furthermore, from table tab02, we conclude that $S \to x_1 \to x_2$ violates Condition 1 or Condition 2 for any x_1 and x_2.

Therefore, we may assume that any subtrail of T satisfies Condition 3 along with Condition 1 and 2 until the end of this proof.

Condition 3 There is no subcycle of the subtrail of length exactly equal to $2n$.

Now, suppose that T has a subcycle

$$S = s_1 \to s_2 \to \cdots \to s_{2n-2} \to s_1$$

of length $2n - 2$. We may express the vertex classes of $K_{n,n}$ as $E = \{s_2, s_4, \ldots, s_{2n}\}$ and $O = \{s_1, s_3, \ldots, s_{2n-1}\}$, where s_{2n-1} and s_{2n} denote the vertices that S does not include. Let x_1, x_2, x_3, and x_4 denote the vertices defined by

$$T = \cdots \to S \to x_1 \to x_2 \to x_3 \to x_4 \to \cdots .$$

By definition, we have $\{x_1, x_3\} \subseteq E$ and $\{x_2, x_4\} \subseteq O$. Then, $x_1 = s_{2n}$ follows from table 3. Then, $x_2 \in \{s_3, s_{2n-1}\}$ follows from table 4. Since $x_2 \neq s_3$ follows from table 5, we have $x_2 = s_{2n-1}$. Furthermore, $x_3 = s_4$ follows from table 6. From table 7, we conclude that $S \to x_1 \to x_2 \to x_3 \to x_4$ violates Condition 1, 2 or 3 for any x_4, a contradiction derived.

We have thus proved the theorem.

□

Table 4. Violations of the conditions when $x_1 = s_{2n}$ and $x_2 \notin \{s_3, s_{2n-1}\}$

z	Walk $S \to s_{2n} \to z$ violates ...
s_1	Condition 1
$s_5, s_7, \ldots, s_{2n-3}$	Condition 2

Table 5. Violations of the conditions when $x_1 = s_{2n}$ and $x_2 = s_3$

z	Walk $S \to s_{2n} \to s_3 \to z$ violates ...
s_2, s_4, s_{2n}	Condition 1
$s_6, s_8, \ldots, s_{2n-2}$	Condition 2

Table 6. Violations of the conditions when $x_1 = s_{2n}$, $x_2 = s_{2n-1}$, and $x_3 \neq s_4$

z	Walk $S \to s_{2n} \to s_{2n-1} \to z$ violates ...
s_{2n}	Condition 1
$s_6, s_8, \ldots, s_{2n-2}$	Condition 2
s_2	Condition 3

Table 7. Violations of the conditions $x_1 = s_{2n}$, $x_2 = s_{2n-1}$, and $x_3 = s_4$

z	Walk $S \to s_{2n} \to s_{2n-1} \to s_4 \to z$ violates ...
s_3, s_5, s_{2n-1}	Condition 1
s_1	Condition 2
$s_7, s_9, \ldots, s_{2n-3}$	Condition 2

The following lemma and proof are both presented in our article[4]. The proof in the previous article, however, includes many errors.

Lemma 2 *Let m and n be positive integers with $n < m$. Then,* $\mathrm{ERL}(K_{2m,2n}) = 4n$ *holds.*

Proof. Let $A = \{(0,x) \mid 0 \leq x \leq 2n-1\}$ and $B = \{(1,x) \mid 0 \leq x \leq 2m-1\}$ be the vertex classes of the complete bipartite graph $K_{2m,2n}$. Let k denote positive integer $\gcd(m,n) = \gcd(2m,2n)/2$. For each $j \in \{0,1,\ldots,k-1\}$, we define trail H_j and circuit C_j of G as follows:

$$\begin{aligned}
H_j &= (0,0) \to (1,2j) \to (0,1) \to (1,(2j+1) \bmod 2m) \to \cdots \to \\
&\quad (0, i \bmod 2n) \to (1,(2j+i) \bmod 2m) \to \cdots \to (0,2n-1) \to (1,(2j-1) \bmod 2m), \\
C_j &= H_j \quad (0,0).
\end{aligned}$$

That is to say, $(0,0)$ is the initial vertex of H_j and C_j, and for every $i \in \{0,1,2,\ldots,2n-1\}$, $(0, i \bmod 2n)$ and $(1,(2j+i) \bmod 2m)$ are $2i$ and $2i+1$ edges distant from $(0,0)$ on C_j, respectively. Furthermore, the length of C_j is the double of the minimum of positive integer i that satisfies both

$$i \equiv 0 \pmod{2n} \quad \text{and} \quad 2j+i \equiv 2j \pmod{2m}\,. \tag{1}$$

Condition (1) is equivalent to the following:

$$i \text{ is a multiple of } 2k, \quad i/(2k) \equiv 0 \pmod{n/k}, \quad \text{and} \quad i/(2k) \equiv 0 \pmod{m/k}\,,$$

and, by the Chinese remainder theorem, there is a unique integer i that satisfies the condition above and $0 < i/(2k) \leq nm/k^2$. Since $i = 2nm/k$ satisfies condition (1), the length of C_j is $4nm/k$ for every j.

For any integers i and j with

$$0 \leq i \leq k-1, \quad 0 \leq j \leq k-1, \quad \text{and} \quad i \neq j, \tag{2}$$

there is no edge e such that $e \in E(C_i) \cap E(C_j)$, where $E(C_i)$ and $E(C_j)$ denote the set of all the edges on C_i and C_j, respectively. It is for the reason that a contradiction follows from the existence of such an edge e as follows. Assume that there is such an edge e. Then, there must be integers p and q such that one of the following three conditions holds:

(a) $(0, p \bmod 2n) = (0, q \bmod 2n)$ and $(1,(2i+p) \bmod 2m) = (1,(2j+q) \bmod 2m)$.

(b) $(0, p \bmod 2n) = (0, q \bmod 2n)$ and $(1,(2i+p) \bmod 2m) = (1,(2j+q-1) \bmod 2m)$.

(c) $(0, p \bmod 2n) = (0, q \bmod 2n)$ and $(1,(2i+p-1) \bmod 2m) = (1,(2j+q-1) \bmod 2m)$.

It is impossible that condition (b) holds since $p \bmod 2n = q \bmod 2n$ implies $p \equiv q \pmod 2$, and $(2i+p) \bmod 2m = (2j+q-1) \bmod 2m$ implies $p \not\equiv q \pmod 2$. It is also impossible that condition (a) or (b) holds. If condition (a) or (b) holds, then both $p - q \equiv 0 \pmod{2k}$ and $i - j \equiv 0 \pmod k$ hold. Then, $i - j \equiv 0 \pmod k$ contradicts condition (2). Hence, the circuit

$$T = H_0 \to H_1 \to \cdots \to H_{k-1} \to 0$$

obtained by connecting the circuits C_0, C_1, $\ldots$, C_{k-1} in this order is an Eulerian circuit of $K_{2m,2n}$.

We now can readily verify the following two facts. Each vertex in A and B appears at regular $4n$ and $4m$ edges intervals, respectively, on C_j for each $j \in \{0,1,2,\ldots,k-1\}$. Furthermore, if a vertex v appears at position p in trail H_j and q in trail $H_{(j+1) \bmod k}$, then p and q are at least $4(m-1) \geq 4n$ edges distant each other. Thus, we conclude the proof. □

4. The Eulerian recurrent lengths of complete graphs

We give an upper and lower bound on the ERL of complete graphs K_n that consists of odd number of vertices in this section.

4.1. An upper bound on the ERL's of complete graphs

We give an upper bound on the ERL's of complete graphs K_n for odd integers n greater than or equal to 5 as follows:

$$\mathrm{ERL}(K_n) \leq n-2\,,$$

which immediately follows from the following theorem. The theorem and the proof are both presented in [3].

Table 8. Violations of the conditions $x_1 = s_{2n}$, $x_2 = s_{2n-1}$, and $x_3 = s_4$

z	Walk $S \to z$ violates ...
s_1	Condition 4
s_2, s_n	Condition 5
$s_4, s_5, \ldots, s_{n-1}$	Condition 6

Table 9. Violations of the conditions $x_1 = s_{2n}$, $x_2 = s_{2n-1}$, and $x_3 = s_4$

z	Walk $S \to s_3 \to z$ violates ...
s_3	Condition 4
s_1, s_2, s_4	Condition 5
$s_5, s_6, \ldots, s_n$	Condition 6

Theorem 2 *Let n be an odd integer with $n \geq 5$. Then, every Eulerian circuit of K_n has a subcycle of length at most $n-2$.*

Proof. The strategy for the proof is similar to that of Theorem 1. Let T be an Eulerian circuit of K_n. We derive a contradiction from the assumption that the length of a shortest subcycle of T is greater than $n-2$, proving that T always has a subcycle of length at most $n-2$. Any subtrail S of T always satisfies the following three conditions.

Condition 4 S has no loops, where a loop is an edge joining a vertex to itself.

Condition 5 For any pair of vertices (v, w), S does not have two or more edges joining v and w

Condition 6 S has no subcycles of length at most $n-2$.

Since the order of K_n is n, T has a subcycle S of length at most n. First, suppose that T has has a subcycle

$$S = s_1 \to s_2 \to \cdots \to s_n \to s_1$$

of length n. Let x_1 and x_2 be the vertices such that

$$T = \cdots \to S \to x_1 \to x_2 \to \cdots .$$

Then, $x_1 = s_3$ follows from table 8. Furthermore, it follows from table 9 that $S \to x_1 \to x_2$ violates Condition 4, Condition 5 or Condition 6 for any x_2. Thus, we obtain a contradiction.

Next, suppose that T has a subcycle

$$S = s_1 \to s_2 \to \cdots \to s_{n-1} \to s_1$$

of length exactly $n-1$. Let s_n denote the unique vertex not contained in S, and x_1, x_2, and x_3 be the vertices such that

$$T = \cdots \to S \to x_1 \to x_2 \to x_3 \to \cdots .$$

Then, we may assume that any subtrail S of T satisfies Condition 7 along with Condition 4, 5 and 6 until the end of this proof.

Table 10. Violations of the conditions $x_1 = s_{2n}$, $x_2 = s_{2n-1}$, and $x_3 = s_4$

z	Walk $S \to z$ violates ...
s_1	Condition 4
s_2, s_{n-1}	Condition 5
$s_3, s_4, \ldots, s_{n-2}$	Condition 6

Table 11. Violations of the conditions $x_1 = s_{2n}$, $x_2 = s_{2n-1}$, and $x_3 = s_4$

z	Walk $S \to s_n \to z$ violates ...
s_n	Condition 4
s_1	Condition 5
$s_4, s_5, \ldots, s_{n-1}$	Condition 6
s_2	Condition 7

Table 12. Violations of the conditions $x_1 = s_{2n}$, $x_2 = s_{2n-1}$, and $x_3 = s_4$

z	Walk $S \to s_n \to s_3 \to z$ violates ...
s_3	Condition 4
s_2, s_4, s_n	Condition 5
s_1	Condition 6
$s_5, s_6, \ldots, s_{n-1}$	Condition 6

Condition 7 There is no subcycle of S of length n.

Then, $x_1 = s_n$ follows from table 10. Furthermore, $x_2 = s_3$ follows from table 11. From table 12, we conclude that $S \to x_1 \to x_2 \to x_3$ violates Condition 4, 5, 6 or 7 for any x_1, x_2, and x_3.

We have thus proved the theorem.

□

4.2. A lower bound on the ERL's of complete graphs

We give a lower bound of the ERL of complete graphs K_n for odd integers n greater than or equal to 7:

$$\mathrm{ERL}(K_n) \geq n - 4 ,$$

which immediately follows from the following theorem stated in our recent work[5]. We obtain the theorem by slightly improving our previous result in [3].

Theorem 3 *Let n be an odd integer with $n \geq 7$. Then, there is an Eulerian circuit C of K_n such that the length of any subcycle of C is greater than or equal to $n - 4$.*

Proof. Assume that the vertex set of complete graph K_n that consists of n vertices is $\{0, 1, 2, \ldots, n-1\}$. Let H_k denote the Hamiltonian path $n-1 \to v_0(k) \to v_1(k) \to \cdots \to v_{n-2}(k)$

of K_n for each $k \in \{0, 1, 2, \ldots, n-2\}$, where $v_i(k)$ is defined recursively as follows:

$$v_i(k) = \begin{cases} k & \text{if } i = 0, \\ (v_{i-1}(k) + i) \bmod (n-1) & \text{if } i > 0 \text{ and } i \text{ is odd}, \\ (v_{i-1}(k) - i) \bmod (n-1) & \text{otherwise.} \end{cases}$$

It is known that every complete graph K_n consisting of odd number k of vertices is decomposed into $(n-1)/2$ Hamiltonian cycles H_0, H_1, ..., $H_{(n-3)/2}$[6]. Figure 2 depicts H_0, H_1, and H_2 for complete graph K_9 consisting of 9 vertices.

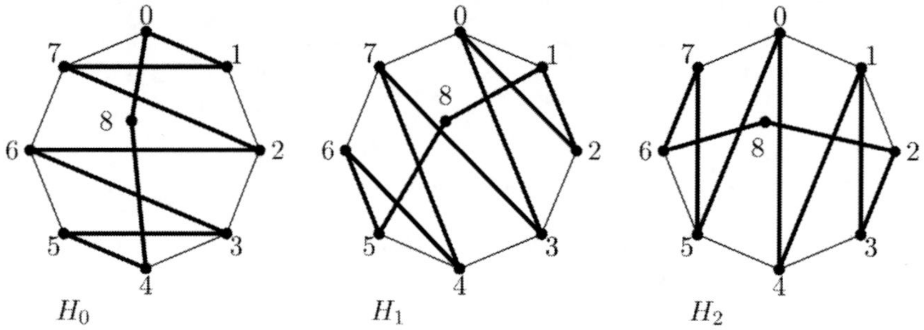

Figure 2. $H_0 \to n-1$, $H_1 \to n-1$, and $H_2 \to n-1$ for K_9.

Let the Eulerian circuit C_n of K_n be defined as

$$C_n = \begin{cases} H_0 \to H_2 \to \cdots \to H_{(n-5)/2} \to H_{(n-3)/2} \\ \quad \to H_{(n-3)/2-2} \to \cdots \to H_1 \to n-1 & \text{if } n \equiv 1 \pmod 4, \\ H_0 \to H_2 \to \cdots \to H_{(n-3)/2} \to H_{(n-5)/2} \\ \quad \to H_{(n-5)/2-2} \to \cdots \to H_1 \to n-1 & \text{if } n \equiv 3 \pmod 4. \end{cases}$$

Then, the following hold.

$$v_i(k) = v_j(k+1) \quad \text{implies} \quad n-2 \le n+j-i \le n+2 \quad \text{for each} \quad k \in \{0, 1, \ldots, (n-5)/2\}\,, \quad (3)$$

$$v_i(k) = v_j(k-1) \quad \text{implies} \quad n-2 \le n+j-i \le n+2 \quad \text{for each} \quad k \in \{1, 2, \ldots, (n-3)/2\}\,, \quad (4)$$

$$v_i(k) = v_j(k+2) \quad \text{implies} \quad n-4 \le n+j-i \le n+4 \quad \text{for each} \quad k \in \{0, 1, \ldots, (n-7)/2\}\,, \quad (5)$$

and

$$v_i(k) = v_j(k-2) \quad \text{implies} \quad n-4 \le n+j-i \le n+4 \quad \text{for each} \quad k \in \{2, 3, \ldots, (n-3)/2\}\,. \quad (6)$$

Thus, it follows from (3), (4), (5), and (6) that the length of any subcycle of C_n is greater than or equal to $n-4$ for any integer n greater than or equal to 7. Thus, we conclude the proof. □

5. Concluding remarks

We have given the exact values of the Eulerian recurrent lengths of complete bipartite graphs in Theorem 1 by gathering our results in two articles. We have then described the proof of those values. We also have given an upper and lower bound on the Eulerian recurrent lengths of complete graphs in Theorems 2 and 3, and have described the proofs. We have obtained the lower bound by slightly improving our previous result.

We have already proved that $\mathrm{ERL}(K_n) \leq n - 3$ holds for every odd integer n greater than or equal to 7, in our recent article[5]. It has been verified by computer experiments that $\mathrm{ERL}(K_n) = n - 3$ holds for each integer $n \in \{7, 9, 11, 13\}$. I currently conjecture that $\mathrm{ERL}(K_n) = n - 4$ holds for every odd integer n greater than or equal to 15. If the conjecture holds, then the Eulerian circuit described in the proof of Theorem 3 is an optimal solution for the ERLP constrained to input only complete graphs that consists of odd number n of vertices with $n \geq 15$.

Acknowledgments
I would like to express my gratitude to Professor Masaji Watanabe for his support and encouragement throughout the research.

References
[1] Pevzner P A, Tang H and Waterman M S 2001 *Proceedings of the National Academy of Sciences of the United States of America* **98** 9748–9753
[2] Mori K and Jimbo S 2010 *IPSJ SIG Notes* **2010-AL-128** 1–4 in Japanese
[3] Jimbo S, Inui Y and Hashiguchi K 1999 *RIMS Kôkyûroku* **1106** 25–36 in Japanese
[4] Jimbo S, Hashiguchi K and Han M 2002 *IEICE technical report – Theoretical foundations of Computing* **101** 49–56 in Japanese
[5] Jimbo S and Maruoka A 2014 *IPSJ SIG Notes* **2014-AL-146** 1–7 in Japanese
[6] Bryant D E 2007 *Surveys in Combinatorics 2007* (*London Mathematical Society Lecture Note Series* vol 346) ed Hilton A J W and Talbot J M (Cambridge University Press) chap 3, pp 67–97

Fire retardancy enhancement of unsaturated polyester polymer resin filled with nano and micro particulate oxide additives

M C S Ribeiro[1], S P B Sousa[1], P R O Nóvoa[2], C M Pereira[1] and A J M Ferreira[2]

[1] INEGI - Institute of Mechanical Engineering and Industrial Management, Rua Dr. Roberto Frias, Porto Portugal
[2] FEUP – Faculty of Engineering of University of Porto, Rua Dr. Roberto Frias, Porto Portugal

E-mail: cribeiro@inegi.up.pt

Abstract. In the last years the traditional construction materials, such as wood, glass and steel, have been increasingly replaced by polymer composite materials due to their superior properties. However, this feature has also raised buildings' combustibility fire hazards. Polymer modification with inorganic nanoparticles can be a potential and efficient solution to control matrix flammability without sacrificing other important properties. In this study a new type of unsaturated polyester based composite materials with enhanced fire retardancy are developed, through polymer modification with nano/micro oxide particles and common flame retardants systems. For this purpose, the design of experiments based on Taguchi methodology and analyses of variance were applied. Samples with different material contents and processing parameters resultant from the L9 Taguchi orthogonal array were produced, and their fire properties assessed and quantified by single-flame source and vertical flammability tests. It was found that material and processing parameters have different effects on different properties. Unsaturated polyester composites modified with nano and micro oxide particles showed better fire performance compared to the neat composite improving at least one fire property whatever the nature of the filler. More thorough studies are required in order to improve mix design formulations towards further fire retardancy enhancement.

1. Introduction

Unsaturated polyester composites (UPC) are increasingly replacing traditional materials like wood, glass and metal, in building, furniture, machinery and vehicle components, as well as in other consumer products. However, UPCs are vey susceptible to high temperatures and behave poorly in fire situations [1-3]. This increasing use UPCs and other polymer composite materials has led to a rise of the fire load in buildings resulting in increasing fire hazards. According to statistical data, these materials have contributed to an exponential decrease of 'escape-time' from 17 minutes, in 1977, to 3 minutes, in 2007 [4].

The use of fireproof resins or the incorporation of flame retardant (FR) systems in a polymer matrix partially improves the fire behaviour of composite materials to some extent; however, these solutions usually lead to a reduction in mechanical properties and have a detrimental effect on composite workability and processing [5]. Due to both high specific surface area and reduced dimension, nanomaterials (NM) reduce the mobility of polymer chains, leading to significant changes on volume, surface and functional properties. Unusual property combinations can be achieved, resulting in a new class of materials for almost every application. Previous research revealed that polymer nanocomposites, with homogeneous nanoparticles dispersion, present improved characteristics over unfilled or micro-filled polymers. Depending on the polymer and nanoparticle nature and content, the main improvements are found in mechanical properties (*e.g.*, toughness and stiffness), and barrier properties such as permeability and solvent resistance. Other interesting results include increased thermal stability and, mainly, the improved ability to promote flame retardancy at very low content levels [6-11]. Therefore, UPC modification with NM can be a potential and effective solution to control flammability of polymers, without significant negative effect on other properties.

As early as the 60's nanosized particles have been investigated for fire retardancy and thermal stability purposes of polymer composites [12]. However, it was after Toyota researchers [13-14] found that the addition of clay to polyamide-6 led to superior mechanical properties and substantial increases on heat distortional temperature that this research subject received significant attention. An exhaustive review of the state of the art in this field can be found in Morgan and Wilkie [7]. The main findings, summarized in the book, as well as ongoing research, clearly show that nanoparticles are in fact the basis for improved FR systems. The majority of research work on the subject of thermal stability and fire retardancy of polymers has been carried out on polymer-clay nanocomposites. Other emerging nanoparticles which have also shown promising effects on polymer thermal degradation are nano-oxides. Most of the few studies focused on this issue show promising results, attesting that nano-oxides incorporation can improve thermal stability and other relevant properties of the final product [10, 15-21]. Synergetic effects between metal nanoparticles, including nano-oxides, and conventional FR additives have also been reported [5, 22-25], and this approach appears to be very promising towards an effective improvement in the flame retardancy of UPCs.

Under this framework, the present study is an effort to develop a new UPC material with improved fire reaction behaviour, through matrix modification with hybrid flame retardant systems based on nano-oxides (alumina and magnesium hydroxide), micro-oxides (silica) and phosphinates. Experiments were planned and conducted according to Taguchi methodology with basis on the L9 Taguchi orthogonal array, with 4 material/processing parameters at three variation levels. Fire reaction properties were analysed and quantified by single-flame source and vertical flammability tests.

2. Materials and methods

2.1. *Raw materials and UPC manufacturing*

A commercially available unsaturated polyester resin, with trade name Aropol® FS3992 (Ashland Chemical Hispania S.L., Spain) was used as matrix. Physical and mechanical properties of the cured resin, as supplied by the manufacturer, are displayed in table 1.

Table 1. Physical and mechanical properties of cured resin

Property	Method	Values
Heat distortion temperature (°C)	ASTM D 648	90-100
Water absorption (%)	ASTM D 570	0.2
Tensile strength (MPa)	ASTM D 638	50-70
Flexural strength (MPa)	ASTM D790	90-110
Barcol hardness (-)	ASTM D 2583	45
Ultimate elongation (%)	ASTM D 638	3

The phosphinate based flame retardant was Exolit® OP 1240 (Clariant-Químicos, Lda., Portugal) which consists of a fine white grain-based organic phosphonate powder having high phosphorus content. The silica sub-microparticles (SiO_2, 98% purity) were provided by Innovnano (Portugal). The NanoDur® (Al_2O_3, 99.5% purity, Alfa Aesar®) alumina nanoparticles were purchased from Cymit Quimica S.L. (Spain), and magnesium hydroxide nanoparticles ($Mg(OH)_2$, 99% purity) were supplied by NanoAmor® (USA). All fillers have spherical shape, and some of their physical properties are specified in table 2.

Table 2. Physical characteristics of fillers.

Filler	Specific surface area (m^2/g)	Specific gravity (g/cm^3)	Average particle size (nm)
Al2O3	36	3.60	45
Mg(OH)2	80	2.36	15
SiO2	64	2.69	437
Exolit OP1240	-	1.35	25000-50000

A well determined quantity of filler was manually pre-mixed with the unsaturated polyester resin and accelerator (0.5 wt. %) for a few minutes at room temperature. The mixing process was then completed by ultrasonication (0.5s on/0.5s off) for several minutes. The initiator (1.0 wt. %) was then added and mixed. All the mixtures were poured into a preheated mould (1 h/80°C) and the samples were then allowed to cure (6 h/80°C). After de-moulding they were subject to an additional thermal treatment (3 h/80°C).

2.2. *Trial formulations*

The design of experiments (DoE) based on a Taguchi methodology was applied to the experiments plan. DoE is an approach for systematically varying the controllable input factors and observing the effects of these factors on the output product parameters. Using Taguchi methodology, the same information provided by an experience plan based on full factorial method can be assessed but with the advantage of requiring smaller number of experiments. This method can also identify the factors and possible interactions between factors that most influence the response, using analyses of variance (ANOVA) [26, 27]. The main objective is to determine the optimal formulations of UPCs modified with nano/micro oxides and phosphinates that result in the best fire reaction properties. In this study the selected control material and processing factors were: type of oxide particle (A) and content (B), phosphinate FR content (C) and total sonication time (D). Each of these factors was run at three variation levels as shown in table 3. The 4^3 full factorial design (4 factors at 3 variations levels) leads to 64 different experiments; however, using a suitable Taguchi orthogonal array, the experimental plan dimension can be significantly reduced. In the present case the L9 orthogonal array was selected, which results in 9 different experiments. The resultant trial formulations are specified in table 4.

Table 3. Design factors and levels.

Levels	Factor A	Factor B (wt%)	Factor C (wt%)	Factor D (min)
1	$Mg(OH)_2$	0.0	0.0	3
2	Al_2O_3	2.5	10.0	15
3	SiO_2	5.0	15.0	9

Table 4. Trial formulations resultant from L9 Taguchi array.

Trials	Factor A	Factor B	Factor C	Factor D
F1	$Mg(OH)_2$	0.0	0.0	3
F2	$Mg(OH)_2$	2.5	10.0	15
F3	$Mg(OH)_2$	5.0	15.0	9
F4	Al_2O_3	0.0	10.0	9
F5	Al_2O_3	2.5	15.0	3
F6	Al_2O_3	5.0	0.0	15
F7	SiO_2	0.0	15.0	15
F8	SiO_2	2.5	0.0	9
F9	SiO_2	5.0	10.0	3

IOP Conf. Series: Materials Science and Engineering **58** (2014) 012020 doi:10.1088/1757-899X/58/1/012020

2.3 Fire characterization tests

After UPCs preparation according to trial formulations defined in table 4, the fire properties were assessed by the single-flame source and vertical UL-94 flammability tests performed on two and five specimens, respectively, of each formulation. All specimens were pre-conditioned for 48 hours at 23 °C/50 % RH before being tested.

2.3.1 Single-flame source test. In the single-flame source test carried out according to ISO 11925-2 standard test [28], two test specimens (250x90x10mm^3) were subjected to direct action of a small flame for 30 s on the bottom edge (figure 1). The maximum test duration was 1 minute (60 seconds) after flame removal. The classification criteria were based on observations of whether the flame spread reached the 150 mm length marker on the specimen within a given time, and if the filter paper below the specimen ignited due to flaming debris.

2.3.2 UL94-Vertical flammability test. The UL94 (Underwriters Laboratories, USA) test allows the determination of plastics flammability characteristics [29]. The standard classifies plastics according to their burning behaviour in different orientations (horizontal or vertical burning tests) and thicknesses. The vertical method was used in this study. The specimens (125×13×5 mm^3) were supported vertically and a flame was applied to the bottom of the specimen (figure 2). The flame was applied for 10 s and then removed until flaming stops, at which time the flame was reapplied for another 10 s and then again removed. On the basis of overall results for a sample of five specimens, materials can be classified as V-0, V-1, V-2 or unrated (U).

Figure 1. ULC specimen (Trial 1) during single-flame source test.

Figure 2. ULC specimen (Trial 4) during vertical UL-94 flammability test.

3. Results and discussion

3.1. Data test results

The single-flame source test results (*i.e.*, ignition time -IT_{SF}-, total length of specimen material destroyed –MD,- and total time of the test –TT-), and the vertical UL-94 test results (*i.e.*, ignition time – IT_{UL} -, after flame times -T_1, T_2-, and correspondent UL 94 classification) are presented in table 5 and 6, respectively. Presented values are average values obtained for two and five specimens and respective standard deviations (when applicable).

Globally, as shown in tables 5 and 6, filler addition improved at least one fire response parameter when compared to unfilled formulation F1, either by increasing the time to ignition (IT_{SF} or IT_{SF}), or decreasing both total length of specimen material destroyed (MD) and combustion times (TT or T_1 and T_2). Trial formulations F5 and F7 showed particularly significant improvements, with much lower combustion times, which led to a V-1 classification for the vertical UL94 test.

Table 5. Single-flame source test results.

Trials	IT_{SF} (s)	MD (mm)	TT (s)
F1	1.0±0.0	16.3±11.3	60.0± 0.0
F2	2.0±0.0	24.5± 7.8	60.0± 0.0
F3	1.0±0.0	10.5±12.0	42.5±24.7
F4	2.5±0.0	14.0± 7.1	41.0± 2.8
F5	1.5±0.0	17.0± 1.4	45.0±12.7
F6	2.0±0.0	19.0± 0.0	60.0± 0.0
F7	2.0±0.7	4.0± 0.0	40.0± 0.0
F8	1.5±0.0	25.5± 4.9	60.0± 0.0
F9	2.5±0.7	11.0± 1.4	60.0± 0.0

Table 6. Vertical UL-94 test results.

Trials	IT_{UL} (s)	T_1 (s)	T_2 (s)	Class
F1	2.0±0.0	281.0±24.4	-	U
F2	3.4±0.3	18.6± 4.6	39.2±23.8	U
F3	2.6±0.4	43.0± 4.2	35.5±27.5	U
F4	2.2±0.4	51.0±30.4	23.0±27.7	U
F5	2.0±0.0	2.2± 2.6	29.5±18.0	V-1
F6	2.5±0.9	333.2±46.3	-	U
F7	2.0±0.7	2.8± 2.2	15.0±21.5	V-1
F8	1.9±0.1	229.0±23.2	-	U
F9	2.5±0.2	56.6±29.7	120.0±20.0	U

3.2. Data treatment and analysis of results

The Taguchi method uses ANOVA to analyse data obtained from implementation of the experimental design. The ANOVA performs the overall variance analysis of a sample set, identifying its origins and assessing the contribution of each variable or factor to the global dispersion.

The various parameters obtained from both single-flame source and vertical UL-94 fire tests can be submitted to statistical treatment with two exceptions: T_2 and TT parameters. During the UL-94 test, great part of the specimens were not submitted to the 2nd flame application because they burned completely during and/or after removal of the 1st flame; on the other hand, in the single-flame source test, the total combustion time computed for F1, F2, F6, F8 and F9 corresponded to the maximum time allowed for this test (60 s). Hence, the parameters submitted to statistical treatment were limited to both ignition times (IT_{SF} and IT_{UL}), 1st after flame time (T_1), and total length of material destroyed (MD).

The results of the analyses of variance for the aforementioned parameters are presented in tables 7 and 8. The results are detailed in terms of sum of squares (SQ), degrees of freedom (dF) quadratic mean deviation (QMD), F-ratio (F_{rat}) and percent contribution to global variation (P). All analyses were performed for a significance level (α) of 5% (or a confidence level of 95%). For this level of significance, the F-test ratio critical value (F_c) is 3.26 for all 4 factors under study.

Response graphics regarding the main effects of each factor on the analysed fire properties are displayed in figure 3.

Table 7. Analyses of variance of single-flame source test results (IT_{SF} and MD).

	Variation Source	SQ	dF	QMD	F_{rat}	P (%)
IT_{SF}	A	1.778	2	0.889	4.00	18.8
	B	0.111	2	0.056	0.25	Rej.
	C	2.778	2	1.389	6.25	32.8
	D	0.444	2	0.222	1.00	Rej.
	Er	2.000	9	0.222		53.3[a]
	Tot	7.111	17			100.0
MD	A	63	2	32.8	0.69	Rej.
	B	382	2	190.9	4.17	23.6
	C	353	2	176.7	3.86	21.3
	D	19	2	9.6	0.21	Rej.
	Er	412	9	45.8		63.3[a]
	Tot	1130	17			100.0

Table 8. Analyses of variance of vertical UL-94 flammability test results (IT_{UL} and T_1).

	Variation Source	SQ	dF	QMD	F_{rat}	P (%)
IT_{UL}	A	0.470	2	0.235	1.27	Rej.
	B	0.226	2	0.113	0.61	Rej.
	C	1.448	2	0.724	3.93	10.0
	D	2.072	2	1.036	5.62	15.7
	Er	6.636	36	0.184		74.7[a]
	Tot	10.852	44			100.0
T_1	A	123157	2	61579	19.5	16.0
	B	68526	2	34263	10.8	8.5
	C	396093	2	198046	62.7	53.5
	D	27214	2	13607	4.3	2.9
	Er	113763	36	3160		19.1[a]
	Tot	728752	44			100.0

[a] Residual error computed as 100% minus the sum of percent contributions relative to all factors (A, B, C and D).

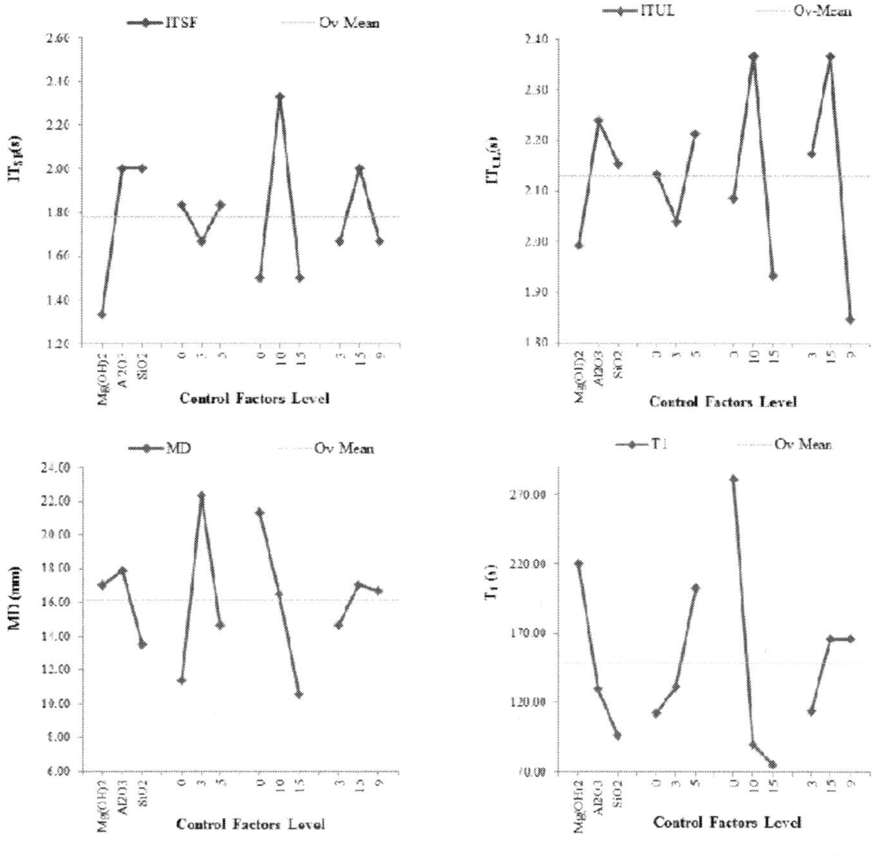

Figure 3. Main effects of material factors on fire reaction properties: time to ignition (IT_{SF} and IT_{UL}), total length of material destroyed (MD) and 1st after flame time (T_1).

As shown in tables 7 and 8, different effects with different relative weights were found for each material factor on the global variation of fire properties under study. The major outputs can be summarized as follows:

a) The oxide particle content (Factor B), for a confidence level of 95%, has no significant statistical effect on global variation of times to ignition regardless of the used fire test. Concerning this target response (*i.e.*, the time to ignition) the null hypothesis is also rejected for the type of oxide particle (Factor A) when the UL-94 flammability test is used. The type of oxide particle, as well as the sonication time (Factor D), have also no relevant effect on how fast the material is destroyed (MD) by fire.

b) Overall based on ANOVA results, it can be stated that the factor with the lowest effect on the analysed fire properties is sonication time (Factor D) whereas the most influential factor is the content of phosphorous based FR (Factor C). For a significance level of 5%, the null hypothesis was rejected for Factor D as regard to MD and IT_{SF} target responses, and the same factor also presented a minor effect (2.9%) on global variation of combustion time (T_1 – UL 94). Although a relatively high value of percent contribution is found for this factor on global variation of IT_{UL} (15%), the high variation due to random error associated to this specific set of experiments (75%) prevents a definite conclusion. On the other hand, Factor C presented the highest percent contributions to global variations of IT_{SF} and T_1 (32.8% and 53.5%, respectively) and it is the second factor with higher influence on IT_{UL} and MD (10% and 21.3, respectively).

Considering the response graphs, the best combinations of factors' levels leading to an improved response for each analysed property are shown in table 9. For each optimal formulation, the estimated response value (R_{EST}) predicted by Taguchi method and determined according to equation (1), is also presented.

$$R_{EST} = A + \Sigma\,(X_i - A) \qquad (1)$$

where A is the global average of all formulations, and X_i represents the marginal mean corresponding to all responses involving factor X with the optimum level (only the material factors with a statistical significant effect on global response are considered).

Table 9. Optimal formulations and estimated response values

Response property	Optimal formulation	Estimated value
IT_{SF}	$A_{2\text{ or }3}\,B_{1\text{ or }3}\,C_2\,D_2$	2.55 s
MD	$A_3\,B_1\,C_3\,D_1$	5.72 mm
IT_{UL}	$A_2\,B_3\,C_2\,D_2$	2.61 s
T_1	$A_3\,B_1\,C_3\,D_1$	0.00 s

4. Conclusions

The Taguchi methodology was used as an alternative approach to the full factorial experimental work plan, to determine material factors' effects on fire properties of polyester composites modified with nano/micro oxides and phosphinates. The following conclusions can be drawn from the investigation:

• It was observed at naked eye that even for the longest used sonication time, filler dispersion was not fully achieved;

• With increasing sonication time a greater content of 'air bubbles' resulted, and the created paths allowing easier flame propagation lead to a poorer burning behaviour;

• The use of nano and/or micro particulate oxide additives improved at least one fire behaviour property. The addition of phosphinate based FR on its own or combined with nano-alumina lead to significant improvements in fire properties, achieving a V-1 rating. This may be an indication of a synergistic effect between the phosphinate compound and nano-alumina.

•The optimized formulations are divergent with respect to some response properties as the factors had different impacts on final results of trial formulations. Nevertheless, it can be stated that 0% to 5% of nano alumina or micro silica particles, in combination with 10% to 15% of phosphinate based FR result in the UPC with overall best fire properties.

The use of nano and/or micro particulate oxide additives may lead to better fire performance than that of unmodified traditional composites, but further studies are required. Additional studies are thus foreseen, to explore the synergistic effect between nano-alumina and conventional FRs, and more thoroughly study the use of these additives (*e.g.*, content, surface treatment), so that products with better fire performance can be obtained.

Acknowledgements

The authors wish to acknowledge FCT, COMPETE and FEDER (under PTDC/ECM/110162/2009 project) for funding the research, Innovnano for donating the SiO_2 micro-particles, and Clariant, for kindly providing the fire retardant Exolit® OP 1240. The authors also wish to thank J. Rodrigues from INEGI for his help and valuable assistance.

References

[1] Ribeiro MCS, Nóvoa PR, Ferreira AJM and Marques AT 2004 Flexural performance of polyester and epoxy polymer mortars under severe thermal conditions *Cement Concrete*

Comp. **26** 803.
[2] Tavares CML, Ribeiro MCS, Ferreira AJM and Guedes RM 2002 Creep behaviour of FRP-reinforced polymer concrete *Compos. Struct.* **57** 47.
[3] Ribeiro MCS, Rodrigues JA, Ferreira AJM and Marques AT 2008 Fire behaviour enhancement of epoxy polymer mortars modified with flame retardant systems *Materials Science Forum* **587-588** 903.
[4] Rakotomalala M, Wagner S and Döring M 2010 Recent developments in halogen free flame retardants for epoxy resins for electrical and electronic applications *Materials* **3** 4300.
[5] Ribeiro MCS, Pereira CMC, Nóvoa PRO, Sousa SPB and Ferreira AJM 2013 Fire reaction and mechanical performance analyses of polymer concrete materials with micro and nano alumina particles *Restoration of Buildings and Monuments* **19** 195.
[6] Mouritz AP and Gibson AG 2006 *Fire Properties of Polymer Composites Materials* (Dordrecht: Springer Publ.)
[7] Manias E, Polizos G, Nakajima H and Heidecker MJ 2007 Fundamentals of polymer nanocomposite technology *Flame Retardant Polymer Nanocomposites* ed AB Morgan and C Wilkie (New Jersey: John Wiley & Sons, Inc.) chapter 2 p 31.
[8] Pereira CMC, Nóvoa P and Marques AT 2008 Dynamical mechanical thermal analysis of flame-retardant unsaturated polyester nanocomposite *Materials Science Forum* **587-588** 232.
[9] Pereira CMC, Herrero M, Labajos F and Rives V 2009 Preparation and properties of new flame retardant unsaturated polyester nanocomposites based on layered double hydroxides *Polym. Degrad. Stabil.* **94** 939.
[10] Tibiletti L, Longuet C, Ferry L, Coutelen P, Mas A, Robin J-J and Lopez-Cuesta J-M 2011 Thermal degradation and fire behaviour of unsaturated polyesters filled with metallic oxides *Polym. Degrad. Stabil.* **96** 67.
[11] Katsoulis C, Kandare E and Kandola BK 2011 The combined effect of epoxy nanocomposites and phosphorus flame retardant additives on thermal and fire reaction properties of fiber-reinforced composites *J. Fire Sci.* **29** 361.
[12] Friedlander HZ 1964, Organized polymerization III: Monomers intercalated in montmorillonite *J. Appl. Sci.* B **2** 475.
[13] Kojima Y, Usuki A, Kawasumi M, Okada A, Fukushima Y, Karauchi K and Kamigaito O 1993 Synthesis of nylon 6-clay hybryd *J. Mater. Res.* **8** 1179.
[14] Kojima Y, Usuki A, Kawasumi M, Okada A, Fukushima Y, Karauchi K and Kamigaito O 1993 Mechanics properties of nylon 6-clay hybrid *J. Mater. Res.* **8** 1185.
[15] Laachachi A, Ferriol M, Cochez M, Cuesta J-ML and Ruch D 2009 A comparison of the role of boehmite (AlOOH) and alumina (Al2O3) in the thermal stability and flammability of poly(methyl methacrylate) *Polym. Degrad. Stabil.* **94** 1373.
[16] Laachachi A, Cochez M, Leroy E, Gaudon P, Ferriol M and Lopez-Cuesta JM 2006 Effect of Al2O3 and TiO2 nanoparticles and APP on the thermal stability and flame retardance of PMMA. *Polym, Advan. Technol.* **17** 327.
[17] Baskaran R, Sarojadevi M and Vijayakumar C 2011 Unsaturated polyester nanocomposites filled with nano alumina *J. Mater. Sci* **46** 4864.
[18] Zhang M and Singh RP 2004 Reinforcement of unsaturated polyester by aluminum and aluminum oxide nanofillers *Mater. Lett.* **58** 408.
[19] Moreira DC, Sphaier LA, Reis JML and Nunes LCS 2011 Experimental investigation of heat conduction in polyester-Al2O3 and polyester-CuO nanocomposites *Exp. Therm. Fluid Sci.* **35** 1458.
[20] Chen Y, Zhou S, Chen G and Wu L 2005 Functionalization of silica nanoparticles for polypropylene nanocomposite applications *Prog. Org. Coat.* **54** 120.

[21] Kashiwagi T 2007 Progress in flammability studies of nanocomposites with new type of nanoparticles *Flame Retardant Polymer Nanocomposites* ed AB Morgan and C Wilkie (New Jersey: John Wiley & Sons, Inc.) chapter 10 p 285.

[22] Laachachi A, Cochez M, Leroy E, Ferriol M and Lopez-Cuesta JM 2007 Fire retardant systems in poly(methyl methacrylate): Interactions between metal oxide nanoparticles and phosphinates *Polym. Degrad. Stabil.* **92** 61.

[23] Ilia G and Drehe M 2010 Grafted 2-chloroethylphosphonic acid on inorganic supports used as flame retardant for unsaturated polyester resins *Fire Mater.* **34** 271.

[24] Doğan M and Bayramlı E 2014 The flame retardant effect of aluminum phosphinate in combination with zinc borate, borophosphate, and nanoclay in polyamide-6 *Fire Mater.* **38** 92.

[25] Bourbigot S, Samyn F, Turf T and Duquesne S 2010 Nanomorphology and reaction to fire of polyurethane and polyamide nanocomposites containing flame retardants *Polym. Degrad. Stabil.* **95** 320.

[26] Anderson MJ and Whitcomb PJ 2007 Design of Experiments *Kirk-Othmer Encyclopedia of Chemical Technology* 27 Volume Set (New Jersey: John Wiley & Sons, Inc.) pp 1-22.

[27] Ahmad KZK, Ahmad SH, Tarawneh MA and Apte PR 2012 Evaluation of mechanical properties of epoxy/nanoclay/multi-walled carbon nanotube nanocomposites using Taguchi method. *Procedia Chem.* **4** 80.

[28] ISO 11925-2 2010 Reaction to fire tests - Ignitability of building products subjected to direct impingement of flame - Part 2: Single-flame source test *International Standard Organization.*

[29] UL-94 2001 Standard Test for Flammability of Plastic Materials for Parts in Devices and Appliances *Underwriters Laboratories Inc.*

A Model-inspired Phenomenology Constitutive Equation for the Temperature-dependence of Flow Stress at Confined Dimension II

Rahmat Saptono Duryat[1,2] and Choong-Un Kim[1]

[1]Materials Science and Engineering Department, UT Arlington, Arlington TX, USA
[2]Departemen Teknik Metalurgi dan Material Universitas Indonesia, Depok, Indonesia

E-mail: rahmat.duryat@mavs.uta.edu, choongun@uta.edu

Abstract. This paper presents further development of the constitutive equations based on model-inspired phenomenological approach. In our previous approach, three figures of merit functions were examined to study the temperature-dependence of yield stress and found two model equations that may work especially well for the small volume materials in the moderate temperature ranges (2<TM/T<5). We have investigated the underlying assumptions behind each model to substantiate their physical correctness. We examined the parameters in each models using the frame of flow stress being dislocation-mediated, thermally activated, and obstacle-controlled process. These examinations led us to find refined model equations with parameters that can be related to the characteristics of flow mechanisms.

1. Introduction

A physical mechanistic foundation for a dislocation-mediated, thermally activated, obstacle controlled model discussed in [1] can be generally expressed in (1).

$$\sigma(T) = C + B \cdot A(\dot{\varepsilon}, T) \qquad (1)$$

C = *a* constant parameter related to the a-thermal mechanism of dislocation movement
B = *a* constant parameter related to the internal state parameter
$A(\dot{\varepsilon}, T)$ = an Arrhenius type parameter related to the thermal activation of dislocation movement.

Three proposed models, which had been reconstructed from the first principles and established theories, were evaluated for the case of small volume Cu[1]. Two basic models were recommended to study the behavior. In alignment with (1), the first model that was inspired by the temperature-dependence of elastic modulus, can be re-written as (2):

$$\sigma(T) = \sigma_a(\mu(T)) + \hat{\sigma}\frac{1}{T} \qquad (2)$$

$\hat{\sigma}$ = a threshold stress in the initial frame of reference.

In association with (1), another recommended model that was ignited by a well-known Arrhenius[2] and Claysius-Clayperon[3] equations, can be rearranged as motivated.

$$\sigma(T) = \sigma_a \mu(T) + \hat{\sigma} \exp \frac{-\Delta G}{kT} \tag{3}$$

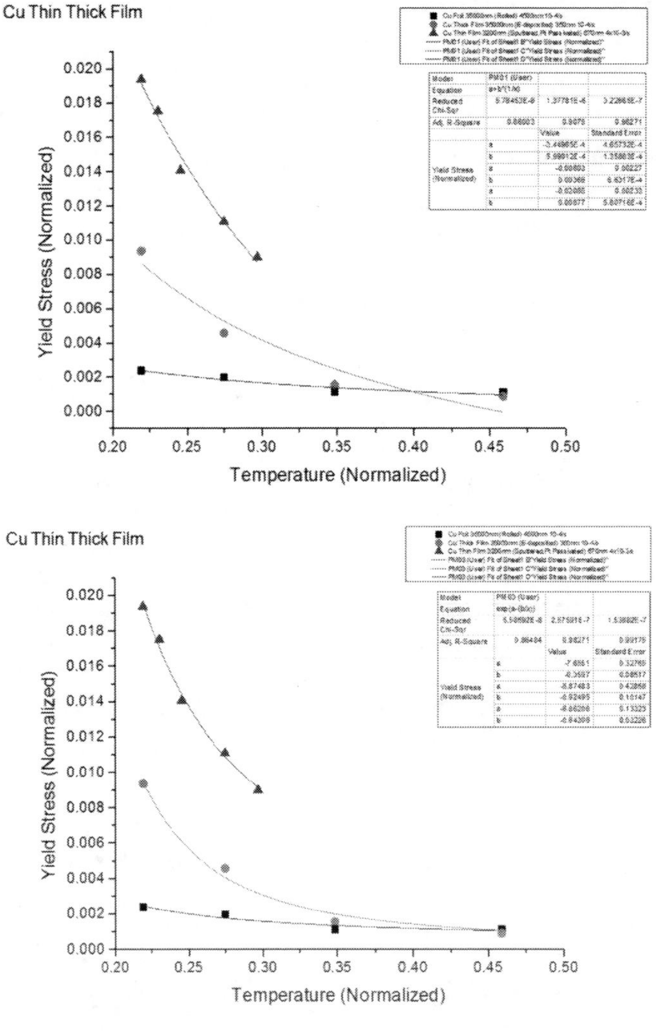

Figure 1. Model I (Top) and Model II (Bottom) to Data Fit in [1] reconfirmed by a non-linear regression analysis [data points from 4,5].

2. Reinterpretation of the Models

In order to get a critical inspection of the models, physical parameters must be able to be readily interpreted from the equation. Re-arranging the variable components on the right hand side of the equations, the model should contain physical parameters that are readily connected to the state of materials. It can be seen that the reciprocal of homologous temperature functions as a scaling factor for the non-linear decrease of yield stress with temperature.

$$\frac{\sigma(T)}{\mu(T)} = \frac{\sigma_a(\mu(T))}{\mu(T)} + \hat{X}\frac{T_M}{T} \tag{4}$$

$\hat{X}$ = a threshold stress in a normalized coordinate, which is assumed to be an internal state parameter.

Equation (4) can be modified into 3 parameters model structure (1), by introducing a constant k_{CUK}

$$\frac{\sigma(T)}{\mu(T)} = \frac{\sigma_a(\mu(T))}{\mu(T)} + \hat{X}\cdot k_{CUK}\frac{T_M}{T} \tag{5}$$

The constant is necessary because it allows the model equation to include the basic idea that the material flow is governed by the probability of successful random attempts (Boltzmann Statistics).

Now, this model allows us to extract the material parameters from the linear plot of normalized strength vs. reciprocal of homologous temperature. Intercept of the line defines a-thermal component of flow stress normalized by the shear modulus. It is generally accepted that this internal stress is related to the interaction of dislocations with long-range barriers.

We reinterpreted (3) by defining a temperature, T_{tact} that represents the characteristic temperature of a thermally activated process. We replace the energy term in the Boltzmann probability function with the temperature so that it can be directly related to the reference temperature (T_M) through a new constant k_{RSD}.

$$T_{tact} = \frac{\Delta G}{k} = \frac{T_M}{k_{RSD}} \tag{6}$$

By doing so, we reserved the physical meaning of the equation as well as the dimensional correctness of the exponential term in the new frame of reference (normalized coordinate system).

$$\frac{\sigma(T)}{\mu(T)} = \frac{\sigma_a(\mu(T))}{\mu(T)} + \hat{X}\exp\frac{-T_M}{k_{RSD}T} \tag{7}$$

Moreover, as it was discussed in the first paper[1], physical parameters could not readily be obtained from a linearized Arrhenius type equation due to the strong correlation between two parameters. Also, the plot of $\ln\frac{\sigma(T)}{\mu(T)}\ vs\ \frac{T_M}{T}$ could have multi-interpretation. To discriminate the parameter and avoid multi-interpretation, we re-arranged (7) prior to linearization.

$$\ln\left[\frac{\sigma(T)}{\mu(T)} - \frac{\sigma_a(\mu(T))}{\mu(T)}\right] = \ln\hat{X} - \frac{1}{k_{RSD}}\frac{T_M}{T} \tag{8}$$

Provided $\frac{\sigma_a(\mu(T))}{\mu(T)}$ is known, k_{RSD} can be obtained directly from the slope(b) of $y=a+bx$ line.

3. Results

Assuming (4) and (6) are nested models, physically sound parameters are depicted in table 1.

Table 1. Physically Sound Parameters obtained from (4), (5) and (8)

Material	Normalized A-thermal Stress Component $\sigma_a(\mu(T))/\mu(T)$ (4,5)	Estimated Normalized Internal State Parameter $k_{CUK}\hat{X}$ (5)	Estimated Normalized Internal State Parameter $\hat{X}$ (8)	Estimated Normalized Energy of Activation $-1/k_{RSD}$ (8)
Cu Rolled Thick Film	-0.0003±0.0005	0.0006	0.0007	0.3085±0.0769
Cu ED Thick Film	-0.0080 ±0.0023	0.0037	0.0044	0.2924±0.0402
Cu SD Pt Passivated Thin Film	-0.0209± 0.0023	0.0088	0.0120	0.2528±0.0127

4. Discussion

Our simple models assumed 3 key parameters. The parameters in these basic models can be physically discriminated into those related to the *long-range order* and *short-range order* stress fields. The entire mechanism that produced a macroscopic plastic deformation is assumed to be driven by the sum of thermally-activated processes of dislocation lines overcoming local obstacles. The *short-range dislocation depinning* can be considered as *a critical gate* for the *long-range dislocation gliding, a conditio sine qua non* for noticeable permanent deformation. Alternatively, the local event can also be thought as *the weakest chain* that resists the entire glide. This gives an idea on how a local mechanism drives a general behavior.

Two *key parameters* of the *local mechanisms* are explicitly proposed in the foundation for the model involving (1) an *internal state parameter* (dislocation spacing) and (2) *a depinning factor* containing the Arrhenius type *rate equations* (Boltzmann probability of successful event and the attempt frequency). While the more complex model (8) explicitly contained both of them, the simpler model (4) has only one explicitly defined.

Despite of its simplicity, equation (4) is superior in defining the plateau stress as it can be exactly obtained from the intercept of the curve while also predicting the internal state parameters. One of the major drawbacks of this basic equation is that it has not discriminated the two parameters related to the local mechanism. We had introduced a constant k_{CUK} to specify the state parameter more accurately. Equation (8), on the other hand, has exactly discriminated the two parameters of local mechanism. This, however, can only be done if the plateau stresses are well defined. In term of practical use, two models are complimentary in giving more accurate estimated parameters.

It can be seen from the results in table 1 that the *normalized internal state parameters* assumed from (5) and (8) is similar in one order of magnitude. These facts indicate that our initial assumption is not unacceptable. It can be more confidently assumed that (4) and (8) are *nested models*. The most important disparity between two models is that in (4) the internal parameter is simply attenuated by the reciprocal of normalized temperature whereas in (6), it is attenuated by an exponential term involving Boltzmann probability function. The drawback of (4) is that it does not reflect explicitly the Boltzmann function. To give a physical meaning of the parameter we introduced a constant that can be physically related to that in (8).

$$k_{CUK} = \frac{T}{T_M}\exp\left(\frac{T_M}{T}\frac{-1}{k_{RSD}}\right) \quad (9)$$

Not only did our new constant k_{RSD} relate the melting temperature T_M to physically sounder temperature T_{tact}, it also eliminated the dimensional problem of using conventional energy of activation parameter terms ΔG with normalized independent variables.

Non local parameter in the basic structure is related to the interaction of gliding dislocation with long-range internal stress field. Although the normalized parameter values of the bottom stress were readily obtained from (4), they are not soundly self-explained. The values of plateau derived from (4) are all negative and so are those obtained from non-linear regression. These *non-sensible* results might be either reinterpreted statistically as 0 (insignificant parameter) or mechanically as (residual) compressive. Another speculative explanation is that they may be caused by the averaging of micro-mechanisms which involves internal short-range and long-range stress fields whose number and proportion should be significantly altered by dimensional constraints.

It can be seen from table 1 that the standard error for the plateau stress is large for all data points larger than an order of magnitude (the parameter should at least larger than approximately 2.5 times of the absolute error). Large error simply means that this parameter was basically not known quite well. In extreme case, since all linear x terms are multiplied by the coefficient, it is statistically sound to replace it with 0.

Although no single data point used in model to data fit is zero or less, we may assume from the process history of sample and specimen preparation and testing method that a little amount of residual compressive stresses might not be completely removed by the pre-testing procedure.

Also, as had been discussed in the first paper (as re-shown in figure 2), the long-range stress field may at a particular condition have a negative value. Since the macroscopic behavior in small volume materials is only averaged over a small cross section, this might become important.

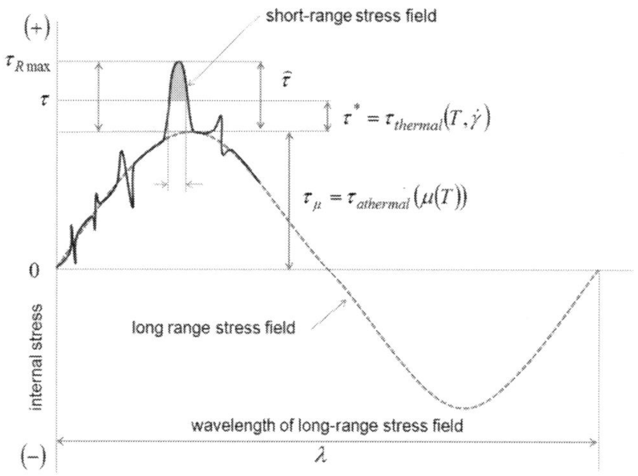

Figure 2. Long-range and short-range stress fields (after[6])

More advanced model, increased data points, and wider range of arguments should hypothetically provide more obvious explanation. This, however, requires a carefully formulation of the merit function. Reinterpreting the two complementary models, a new 3 parameter model representing (5) and (8) as well as analyzing can be written. Global non linear regression analysis of the advanced model had provided more sensible value of plateau stress in addition to the better fit statistics, as indicated by the higher goodness of model to data fitting and level of confidence of the parameters. These are depicted in figure 3, which also illustrate the effect of dimension and process history-structure on key parameters. The inset reflects a common shared mechanism in bulk and small volume materials regardless the characteristics behavior.

Regardless of the shortcomings of the two basics models, they have been able to provide insight into the process and mechanism behind the analysis, which are consistent with our initial assumption. More importantly, assuming them as nested and complementary, physically sound initial parameters can be obtained for more advanced regression analysis.

IOP Conf. Series: Materials Science and Engineering **58** (2014) 012021 doi:10.1088/1757-899X/58/1/012021

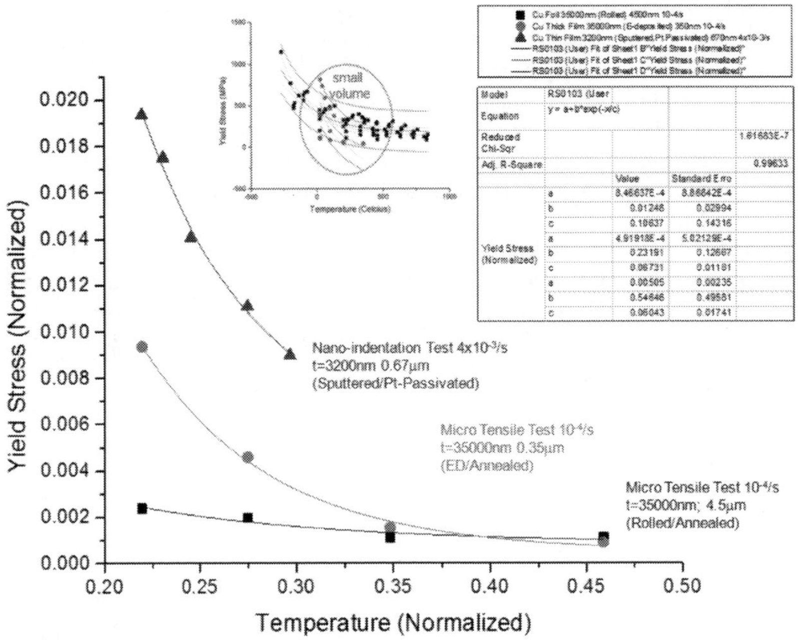

Figure 3. Global Non Linear Regression Analysis of the 3 Parameters Composite (Complimentary) Model (data points from [4,5])

5. Conclusion

Two basic models have been reinterpreted and complimentary employed to reveal physically sound parameters describing a local dislocation-mediated thermally-activated and obstacle controlled process and mechanism. The most important different between these two assumed nested models is that in (4), the internal parameter is scaled directly by the reciprocal homologous temperature whereas in (8), it is scaled by a Boltzmann Probability Function dedicated for the normalized independent variables.

One of the physical parameters that are needed to be reconsidered and reevaluated is the negative value of stress in the obtained parameter from (4). A model of higher parameter needs to be carefully formulated, evaluated, and explored with the aid non-linear regression analysis. Reinterpretation and more advanced regression analysis had resulted in a more plausible value of plateau stress.

The main contribution of the two basic models is to provide an insight into the characteristic mechanism operated in small volume materials and physically sound initial values of parameters. These estimated values are very important in the iterative analysis of more accurate models.

References

[1] Saptono, R and Kim, C-U 2014 *J. App. Mech. Mat.* **534** 83
[2] Arrhenius, S 1889 *Zeit. Phys. Chem.* **4** 226
[3] Frost, A A and Kalwarf, D R 1953 *J. Chem. Phys.* **21** 264-267
[4] Weiss, B, Groger, V, Khatibi, G, Kotas, A, Zimprich, P, Stickler, R, Zagar, B 2002 *Sensors and Actuators A* **99** 172–182
[5] Volinsky, A A, Moody, N R, Gerberich, W W 2004 *J. Mater. Res.* **19** 2650
[6] Conrad, H 1970 *Mater. Sci. Eng.* **6** 265-273

Technological solution for vulnerable communities: How does its approach matter?

CPM Sianipar[1,2,*], K Dowaki[2], G Yudoko[1]

[1] School of Business and Management (SBM), Institut Teknologi Bandung (ITB), Jl. Ganeca 10, Bandung, West Java 40132, INDONESIA
[2] Department of Industrial Administration (IA), Tokyo University of Science (TUS), 2641 Yamazaki, Noda-shi, Chiba-ken 278-8510, JAPAN

* Email: morgana.sianipar@sbm-itb.ac.id; http://morganasianipar.com

Abstract. Vulnerability is an emerging term for both scientific communities and practitioners. Vulnerabilities attributed to a particular community are then aggregated to state it as a vulnerable community. Vulnerability-related study is recognized as an interdisciplinary one due to the complex characteristics of vulnerabilities in each contextual situation. In common understandings, technology is recognized as an interdisciplinary solution, making it possible for supporting any activity in eradicating vulnerability. This study aims to observe such possibilities. Literature survey is taken to investigate the interconnection between vulnerability eradication and technology. Brief comparison between several developing countries and particular focus on Indonesia become a medium of further investigation to reveal issues surrounding technology-related efforts for vulnerability eradication in vulnerable communities. The study reveals that developing countries, particularly Indonesia, tend to adopt approaches from Northern hemisphere, including transferring technologies from developed countries without proper propagation. It means that local knowledge and power are largely ignored in the pursuit of local problem solving for vulnerability eradication. These facts become a signpost to emphasize that approach in implementing technological solution for such purpose is the critical mechanism to ensure the success in every contextual situation. Then, looking at the results of this brief study, its emphasis indicates further requirements to shift the paradigm of typical community development to contextual community empowerment in order to ensure the continuity of every technological solution for a consistent eradication of local vulnerabilities, including possible changes of required approach alongside the shift.

1. Vulnerability: An introduction

Vulnerability has become an important term in today's society. It is used by people across disciplines, including different scientific traditions in seeing and treating a particular phenomenon. From proposals on psychological vulnerabilities to engineering ones, from economics vulnerabilities to ecological ones, the term "vulnerability" has become a contested meaning [1]. Each discipline and scientific tradition has tried to incorporate vulnerability in their own contexts, and has theorized vulnerability as they deal with it from their own perspective in every contextual vulnerable situation. While those disciplines and scientific traditions have proposed their understandings on vulnerability with interconnected meanings, there is no single agreed definition of vulnerability, meaning that the concept of vulnerability itself is rather contextual than general, and diverse rather than converged. The term itself, as a word, is defined as the extent of an observed entity to which it is exposed to potential harm due to its fragile foundation in facing challenges, particularly a shocked one such as natural hazard [2]. The fragile foundation refers to the fundamental factors of an observed entity regarding its capabilities in delivering response, either natural or artificial, for external as well as internal challenges, both natural and artificial. Besides, the exposures are stated as the opportunities of an observed entity to experience possible future harms as the results of either past experiences, present circumstances, or possible future changes. In such understanding, vulnerability itself is the cause of future vulnerabilities as vulnerability is tightly related to fundamental factors of an observed entity, meaning that when a vulnerability produces weak responses to a challenge, the weak responses will accumulate previous vulnerabilities related to the challenge to be bigger vulnerabilities at the future in facing further challenges. Vulnerability, therefore, is highly possible to bring an immediate destruction of an entity [3]. Looking at the diverse definitions of vulnerability, including different approaches in dealing with vulnerability as well as the multi-facets condition of a challenge due to accumulated effects of a vulnerability, understanding vulnerability is rather an interdisciplinary than an isolated thought. Vulnerability can be stated as having multidisciplinary concepts from different perspectives yet almost useless in providing clear answer to overcome a challenge, or incorporated to other concepts as a basis thought in dealing with human-environment interaction [4,5]; however, vulnerability itself is standing as an independent understanding that waits to be approached from all facets of its complexity.

2. Vulnerable communities

2.1. Standing on a fragile foundation

In society, vulnerability is also becoming one of critical concerns in societal development. Recognized as the weak foundation of the extent to which an individual or group of people able to face challenges in surviving their existence, study on vulnerability is particularly focused on community level due to the importance of communal actions in incorporating global movements to regional and local practices [6,7]. Vulnerabilities attributed to a specified community are then aggregately taken to state the community as a vulnerable one. Vulnerability, therefore, is known as a basic characteristic of a specified community in developing and/or recovery contexts. In developing context, vulnerabilities of a vulnerable community refer to exposures experienced by a specified community that cause instability of internal communal system or interactions between members of a community in the development process, meaning that fundamental factors supporting the community or the system have not reached a firm position and cannot support each other when changes of challenges occur due to progress of development (Figure 1). On the other hand, vulnerabilities of a vulnerable community in recovery process refer to the results of a crisis caused by disaster or war/social conflict that affect fundamental stability of a community (Figure 1), including its access to outside world. The instability causes exposures of a community in their process to recover internal condition and its connection to other regions. Instability in a recovery process is highly possible to cause next crises. In the middle of either developing or recovery context, there is an intermediary one which is prevention context. In the context of prevention, vulnerabilities of a developing community are the extent of exposures to which the community that is in a development process is possible to fall into a crisis. Either in a development progress or in recovery process after war/conflicts or natural disasters, including prevention context, vulnerabilities occur as the result of unstable circumstances, an unbalanced state between fundamental factors in a community [8-10], among local/regional entities that construct the foundation of local/regional resilience. Roughly-speaking, at community level vulnerabilities happen as the products of developing and/or recovery contexts, and vulnerable communities exist due to the vulnerabilities deeply-rooted in a community that is critically live in an unstable circumstance caused by continuous treats of natural disasters or unconducive conditions for supporting strong growth and fundamental stability [11-14].

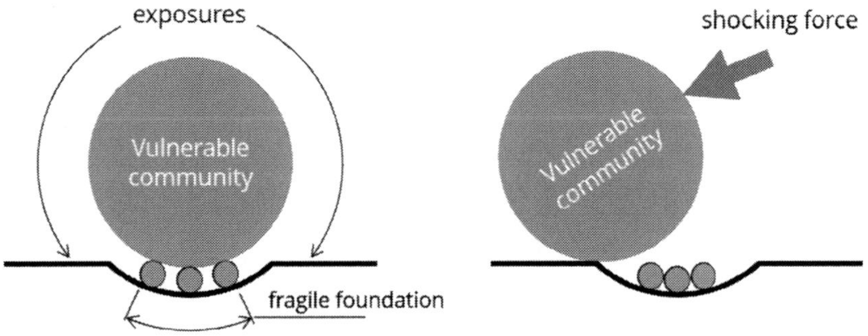

Figure 1. Standing on a fragile foundation.

2.2. *In developing countries: A major part of the world*

Looking at the conception, vulnerable communities arguably exist in both developed and developing countries. As indicated by the understandings of vulnerability, each of developed or developing has its own vulnerabilities that cause the exposure of either national development progress, national recovery process, or national crisis prevention system. Such reason has become the root of widely-accepted argument in which any country can be stated as vulnerable [15-18]. Furthermore, the causes of vulnerabilities can be distinguished to five different conditions [19]: destructive growth, poverty, political rigidity, dependency, and geographic isolation. Those conditions are possible to happen in any community, in any of developing and recovery context, and either in developed or developing countries; however, those indicators in Woodward's study have indicated that vulnerabilities, in other words: vulnerable communities, are most likely to exist in developing countries. Such statement is based on the considerations in which each of those factors strongly exist in developing countries as the result of the developing state and are supported by instability of fundamental factors. While in developed countries fundamental factors of a nation or communities such as economic power or social resilience are firmly founded on solid resilience and can support each other in facing challenges, developing countries still struggle to stabilize their fundamental factors due to fragile foundation. Moreover, in fact 82.5% people all around the world in the year 2013 live in developing countries [20]. Even if China is excluded due to its potential to become a new superpower country, the rests still cover 63.5% of the world population. By looking that those statistics and the highly supported consideration in which vulnerable communities are mostly likely to exist developing countries, therefore, vulnerable communities are one of the world's major entities and a very important facet of the future security of mankind [21-22]. However, the diverse and contextual understandings of vulnerability itself has become a problematic discourse for authorities and/or local bodies of developing countries in coping with vulnerabilities of a vulnerable community in their own area. Fundamental instability also causes more difficulties in eradicating vulnerabilities of local communities. Such barriers in eradicating vulnerabilities of vulnerable communities are worsened by accumulated vulnerabilities that produce other vulnerabilities in the future and are highly possible to bring crisis to local as well as surrounding area. In other words, eradicating vulnerability is a now-or-never deal, meaning that a failure in treating present vulnerabilities will cause much wider exposure in the future; on the contrary, right treatment will close the gate for other vulnerabilities by strengthening and stabilizing fundamental factors of a vulnerable community.

3. Technological solution for vulnerability eradication

3.1. *Technology: An interdisciplinary solution*

Based on above discussions, by considering the interdisciplinary perspectives of vulnerability, instability of fundamental factors of vulnerable community, and wide exposure of developing countries to fall into crisis, vulnerable communities require a cross-disciplinary solution that can support growth, vulnerability eradication, and strengthening local resilience. In that spirit, technology comes as a powerful answer in delivering those purposes in a single solution. Despite the contextual characteristic of vulnerability-related studies, technology itself has been recognized to have critical position as one of the essential concerns throughout the world [23], and has gotten a unique placement as one of the cornerstones of vulnerability eradication in today's society [24]. Particularly since the Industrial Revolution in the late 18th century, technology has changed human perspectives on their ways of doing things [25]. After such phenomenal history, whatever the object is, wherever and whenever people, as individuals or human institutions such as communities, governments, or business, require a booster to their effort in the pursuit of vulnerability eradication and reinforcing resilience, technology would be a common choice among other ones [26,27]. Of course, there are people whose sentiments have been reluctant in supporting those facts. They are ones who state that the belief in the capacity of technology in providing answers for the reduction of vulnerability even if in its smallest form must not be posited as the ultimate and an only reason to put technology above any other counter facts. Those kinds of people have also toughly pointed their finger to technology as the main cause of future human vulnerability problems such as environmental and health [28-30]. However, people with negative perspectives on technology are unable to refuse the fact that, even if technology has many negative effects imposed to surrounding space and environment, the answer of their concerns would be mainly provided by technological advancements [31,32]. The concerns, therefore, have changed to become only a rhetoric in any effort of vulnerability eradication. In practical level, the use of technology in vulnerability eradication is frequently focused on environmental issue as the response or prevention to natural phenomena surrounding a community [33-35], yet later approaches have begun to cover other kinds of vulnerability such as social, political, and/or economic vulnerabilities [1,4,26]. In short, the incorporation of technological advancements in eradicating vulnerabilities has changed the whole movement. Since technology is posited as a comprehensive solution produced from interdisciplinary perspectives, the implementation of vulnerability eradication has become much robust for strengthening fundamental factors of a specified community, and stabilizing cross supports between those factors in supporting the resilience of the community, including the deal with undesirable impacts of technology itself.

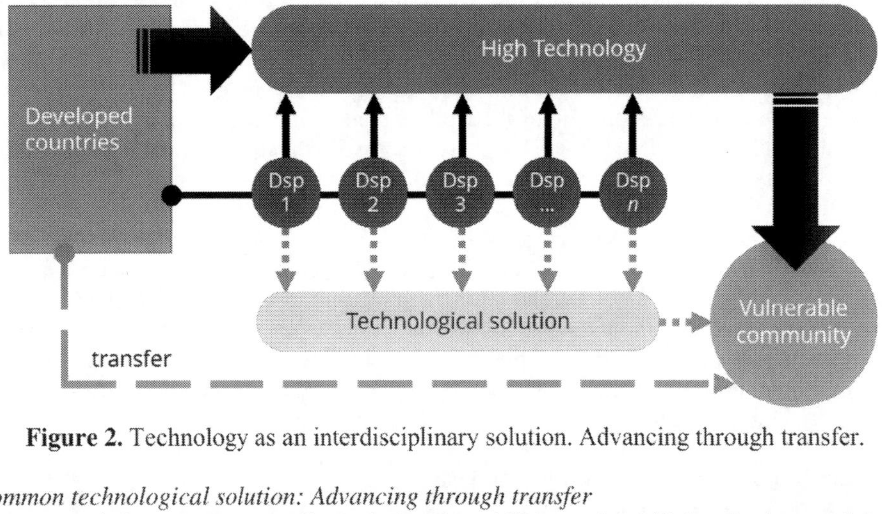

Figure 2. Technology as an interdisciplinary solution. Advancing through transfer.

3.2. Common technological solution: Advancing through transfer

Furthermore, as their attempts to eradicate their vulnerabilities and stabilizing fundamental factors for national resilience, including the pursuit of international recognition to become developed ones, developing countries begin to adapt many approaches from Northern hemisphere in implementing their strategies [36]. Following the terminology transformation of technology from only an artifact to be a more systematic technical enhancement of related processes [23], developing countries have begun to incorporate such understanding into their development strategies. Approaches and solutions for technological advancements in vulnerability eradication, therefore, are adapted from developed countries to enhance practices in developing ones. In that spirit, technological transfer has become a recognized approach to massively bring technologies from developed countries (Figure 2). In technology transfer, technologies, usually ones with high technical specifications and are common kinds used in the Northern hemisphere, are brought from developed countries. In developed countries, a technology is seen as a universal solution for a certain extent of problems; hence each scientific discipline related to the purpose of a technology (Figure 2, Dsp_n) contributes its expertise in developing the technology in order to unifying the specifications of the technology. In that scheme, a technology is supposed to be a multidisciplinary solution, meaning that each scientific discipline becomes a separated part of solutions embedded in the technology. While the technology is brought to be a technological solution for a vulnerable community, the multidisciplinary approach is consequently brought together within it; however, vulnerability eradication is rather cross-disciplinary than multidisciplinary, meaning that every technology from developed world is not suitable enough to eradicate vulnerabilities from all directions simultaneously due to universal technical specifications to cover the extent of problems in some distinctive scientific disciplines to which the technology needs to solve in developed countries. Also, massive technological transfer in order to support vulnerability eradication in vulnerable communities do not practically obliterate all of obstacles in its implementations in developing countries. Even if some alternative concepts such as appropriate technology (AT) [37] and grassroots innovation [38] have been incorporated as the intermediaries in adapting approaches from developed countries, some difficulties are triggered by unsuitable conditions in developing countries that are left behind in many essential aspects compared to the profile of necessary supports in developed countries in which many approaches were first developed.

4. Situational limitations in technology transfer for vulnerability eradication

4.1. Fundamental differences of developed and developing countries

In fact, developed countries have already had strong national resilience as the result of stabilized fundamental factors (Figure 3) such as economic prosperity, social welfare, *etc.*, which then affect the correlation between widely-accepted constructs of national growth and wealth distribution to the choice of technology. Such conditions happen as the results of some applicable assumptions taken in developed countries. For example, on economic side, developed countries don't have much problems in the economies of scale compared to developing ones [39]; or, people in developed countries have low variations of the definition of social capital, meaning that people tend to act correspondingly with the direction of any other citizen in achieving social goals [40]. Although in developed countries there are diversities of conditions and trade-offs, including critical vulnerabilities that are possible to make them fall into crisis, fundamental factors in developed countries would support each other in facing worst challenges, meaning that a weakness in one factor would be overcome by improvements in other ones. Therefore, the steady condition, as stated by Kaplinsky [41], refers to the time when there is a homogeneity of behaviors between fundamental factors of a country. Such perfect condition also affects any technological choice. Due to the overwhelming supports both from government and the civilized society, technological changes are enormously concentrated at capital intensive techniques [42] by interpreting environmental issue through the rate of impacts imposed per contribution to market needs [31]. In that term, technical qualities become the main goal of technological advancements due to accessible resources for reaching any purpose. On the other hand, developing countries arguably exist in different settings. Circumstances and national capabilities in their own situation are not strong enough compared to which developed countries have achieved in order to eradicate vulnerabilities and supporting national resilience [43-45]. Huge varieties (Figure 3), *i.e.* in economic capacities and/or social goals, have decomposed developmental efforts in developing countries into detached entities, from an integrated national strategy into separate works in each targeted area. Such huge disparities have made some approaches focusing on unique conditions in each area to be an essential medium to deal with any existing diversity. Still, the problem in direct adaptation of approaches from developed countries to developing ones, including advances of technology through technological transfer, remains same [46]. Enormous diversities that commonly occur in developing countries are not adequate to do an extensive application of approaches from Northern hemisphere due to different basic mentalities to the original ones [47]. Different with developed ones, developing countries, with all of their limited capabilities, are unable to get adequate control on all of discrete entities throughout their governance in order to ensure vast vulnerability eradication in all of its interdisciplinary focuses. In terms of technological changes, they obsessively attempt to overtake technical capabilities of developed countries by, ironically, neglecting the needs of their own citizens. Also, such fascinating effort is also very dangerous because of huge potential technological disasters [48]. The application of technological advancements for vulnerability eradication, therefore, is affected by those mentalities.

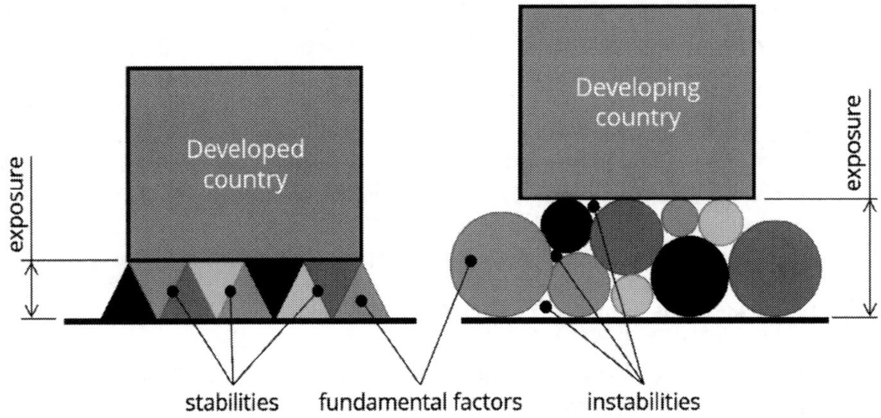

Figure 3. Distinguishing developed and developing countries.

4.2. In brief overview of some developing countries

An example for such condition comes from the world's fourth most populous countries and also the third biggest developing countries: Indonesia. With its 249.5 million population, it covers almost 3.5% of world population [20]. While others big developing countries are continuously showing its substantial growth in conjunction with significant development of its technological advancements for vulnerability eradication, the story of Indonesian remains stagnant due to huge exposures to crisis as the result of fundamental instabilities and social inequality. Countries such as China, India, Brazil, and Mexico have even massively had to interchange their growing potentials to technological transfer for vulnerability eradications both in their own area and in their respective partners of developed countries [49]. In China and India, there is a significant decreasing of dependencies to single directional aids from developed countries due to rapid growth in economic prosperity and technological changes. While China is recognized as new world's economic superpower and followed by its radical advances in technologies, on the contrary, India has its widely-spread technological development from advances to social innovations that strongly affects its ongoing economic growth. Besides, Brazil and Mexico have already had low technical vulnerabilities and good technological capabilities, including growing strategic industries and relatively similar culture to their partners of developed countries, which then produces smooth transfer of technologies despite of social inequalities. On the other hand, technological change for vulnerability eradication in Indonesia is arguably stagnant despite of the growing of industries. In Indonesia, vulnerabilities occur due to both natural disasters and developing state, two conditions that strongly affect many facets of people's life [50-53]. Also, the term vulnerable communities is particularly attributed to people live in rural area, especially for those who work in low-tech industries such as traditional or transitional farming. In fact, of the Indonesian population, more than half live in rural area, and they are largely work as farmers or in water-related jobs. Such kind of communities is then becoming more and more important due to the widening inequalities between urban and rural area. Their massive portion of Indonesian population gives a notion that they have big influences to national competitiveness, meaning that Indonesian authorities have to take opportunities to target any vulnerable community as a means to develop societies' capability in order to reach higher national growths. To do so, some approaches have been proposed. On one side, technological transfer from developed countries is posited as the booster in achieving sustainable growth [54,55]. On the other side, Appropriate Technology (AT), as a promising technological approach to deliver a comprehensive technical solution in a limited condition, is posited to provide a technology with sufficient technical performance at affordable price [53,56].

4.3. The problems arise

However, the practices of those approaches do not reflect the real form of vulnerability eradication. Huge numbers of advanced technologies are purely imported, *i.e.* by Indonesia, from developed countries without proper adaptation, meaning that technologies are taken for granted to pursue rapid economic growth but ignoring indigenous capabilities of local people; besides, AT is becoming important but is implemented through a laboratory-based development process in a workshop behind closed doors with least participation of targeted community members [47,53,55,57] (Figure 4). Those conditions have made technology designers to: often ignore community empowerment, including local context and values; tend to have an exclusion of traditional ways of knowing; have a denial or devaluing people relationships; and strong commitment to industrial (military-like) working styles [58], by which implementation of technological changes and AT in Indonesia is increasingly difficult due to: (1) questionable technological appropriateness; (2) top-down approach to local needs; (3) low technological diffusion; and (4) weak support from local entities [59,60]. Looking at above facts and conceptions, the real problems in the implementation of technological change and AT is not on the vulnerabilities of targeted communities. The problem is the approach consisting methods or techniques to do technological change and developing AT. In current practices, each of them neglects the existence of another, yet field implementation requires intermediation of those approaches in contextual matter as a means to eradicate vulnerabilities and strengthening local resilience through technologies; thus technological solution for vulnerability eradication needs further development of its approach to strengthening the concept and perfecting the practices of vulnerability eradication.

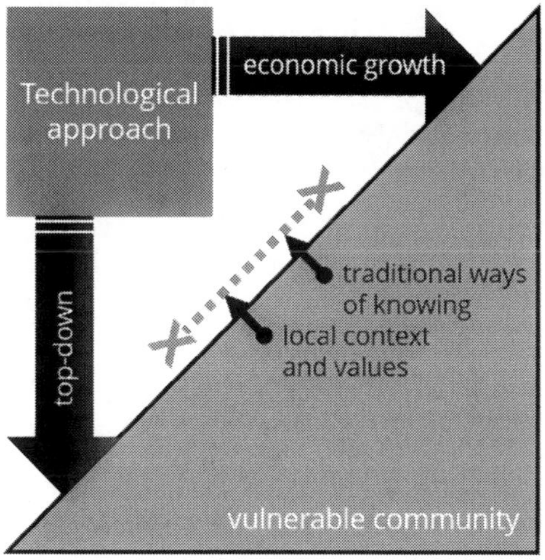

Figure 4. Present technological approach for vulnerability eradication.

IOP Conf. Series: Materials Science and Engineering **58** (2014) 012022 doi:10.1088/1757-899X/58/1/012022

5. Conclusions: The common threads and further researches

Hence, conducting technological advancements and AT in a single timeframe and place needs to incorporate local practices and has to empower local people. Problems discussed above have given notions to intermediate two counterintuitive approaches: technological changes through technology transfer and technology development based on AT approach. Also, there are at least two things that have to be rigorously explored in order to do technological changes for vulnerability eradication. At conceptual level, technological changes for vulnerability eradication requires a clearer understanding to empower people rather than only giving technology to do local development. To do so, scientists need to revisit the concepts of technological changes for vulnerability eradication in vulnerable communities, including an early form of framework to theorize research gap. In terms of its practice, a new framework is required to intermediate the mentality of technology transfer and the spirit of local problem solving; thus technologists could expand the theoretical framework to be a practical framework to conduct technology development for vulnerable communities. Therefore, some pivotal questions such as "what kind of technological changes suitable for vulnerability eradication in vulnerable communities?", "how do anyone can systematically produce a technology for a vulnerable community based on both technological transfer and local problem solving?", and "how to apply such guidance in a field application?" would be very important in order to achieve the success of vulnerability eradication through technological changes at both conceptual and practical levels. The first pivotal question is purposed to deeply investigate the basic model of technological solutions overtime to ensure the continuity and persistency of vulnerability eradication. After that, the second question is supposed to be the initiation of further development of the basic model produced from the first question to construct a new systematic technological approach which combines typical technology transfer and local ways of problem solving. Then, the third question is taken as the basis of testing process of the new systematic technological approach produced based on the second question to investigate the validity and reliability of the new approach directly in field applications. In addition, the intention of bottom-up approach in implementing technological changes for vulnerability eradication in a vulnerable community may also affect the choice of developmental paradigm taken as the basis of any concept and practice. Looking at the previous discussion, there is a need to shift the paradigm from typical development to empowerment. The shift is required to ensure the incorporation of local problem solving to any developmental work, including technology-based vulnerability eradication. Finally, the shift may also further affect some related issues. Empowerment-based technological solution for a vulnerable community may become the entrance point of deeper diffusion of the technology to local process as a means to maintain the continuity of technological changes and the consistency of vulnerability eradication; thus technology acceptance – on community side – and technology suitability – on technology side – have to be carefully taken as the interconnected attributes to push the diffusion further in order to seamlessly integrate the technology to be an integral part of local daily routines.

References

[1] Adger W N 2006 *Global. Environ. Chang.* **16** (3) 268-281
[2] Turner B L, Kasperson R E, Matson P A, McCarthy J J, Corell R W, Christensen L, Eckley N, Kasperson J X, Luers A, Martello M L, Polsky C, Pulsipher A and Schiller A 2003 *P. Natl Acad. Sci. USA.* **100** (14) 8074-8079
[3] Bourdelais P 2005 *Ann. Démogr. Hist.* **110** (2) 11-15
[4] Cutter S L, Boruff B J and Shirley W L 2003 *Soc. Sci. Quart.* **84** (2) 242-261
[5] Turner B L, Matson P A, McCarthy J J, Corell R W, Christensen L, Eckley N, Kasperson J X, Luers A, Martello M L, Polsky C, Pulsipher A, Schiller A, Selin H and Tyler N 2003 *P. Natl Acad. Sci. USA.* **100** (14) 8080-8085
[6] Leichenko R M and O'Brien K L 2002 *Mitig. Adapt. Strat. Global. Chang.* **7** (1) 1-18
[7] Few R 2003 *Prog. Dev. Stud.* **3** (1) 43-58
[8] Brooks N 2003 *Tyndall. Centre. Clim. Chang. Res. Working Paper* **38** 1-16

[9] Hovden J 2004 *J. Risk. Res.* **7** (6) 629-641
[10] Rigg J, Law L, Tan-Mullins M A Y and Grundy-Warr C A R L 2005 *Geogr. J.* **171** (4) 374-379
[11] Comfort L, Wisner B, Cutter S, Pulwarty R, Hewitt K, Oliver-Smith A and Krimgold F 1999 *Environ. Hazard.* **1** (1) 39-44
[12] Graham L R 2006 *Anthropol. News.* **47** (7) 4-5
[13] Eisenman D P, Cordasco K M, Asch S, Golden J F and Glik D 2007 *Am. J. Public. Health.* **97** (Supplement 1) S109-S115
[14] Mechanic D and Tanner J 2007 *Health. Affair.* **26** (5) 1220-1230
[15] Downing T E 1992 *Climate change and vulnerable places: Global food security and country studies in Zimbabwe, Kenya, Senegal and Chile* Available at http://goo.gl/TiU5O0 (Oxford, UK: Environmental Change Unit – University of Oxford)
[16] Adger W N and Vincent K 2005 *C. R. Geosci.* **337** (4) 399-410
[17] Conley T 2009 *The vulnerable country: Australia and the global economy* (Sidney, AU: University of New South Wales Press)
[18] Measey M 2010 *Global. Major. E-J.* **1** (1) 31-45
[19] Woodward A, Hales S and Weinstein P 1998 *Clim. Res.* **11** (1) 31-38
[20] PRB 2014 *World population datasheet 2013* Available at http://goo.gl/vXIY0a (Washington DC, US: Population Reference Bureau)
[21] Morrow B H 1999 *Disasters.* **23** (1) 1-18
[22] Amexo M, Tolhurst R, Barnish G and Bates I 2004 *Lancet.* **364** (9448) 1896-1898
[23] Grübler A 1998 *Technology and global change* (Cambridge, UK: Cambridge University Press)
[24] Füssel H M 2007 *Global. Environ. Chang.* **17** (2) 155-167
[25] Wren D A 2005 *The history of management thought* (New York, US: John Wiley & Sons)
[26] Liverman D 1990 Vulnerability to global environmental change *Understanding Global Environmental Change: The Contributions of Risk Analysis and Management* eds R Kasperson et al (Worcester, US: Clark University) chapter 4 pp 27-44
[27] Willoughby K W 1990 *Technology choice: A critique of the appropriate technology movement* (Colorado, US: Westview Press)
[28] Schumacher E F 1973 *Small is beautiful: Economics as if people mattered* (Virginia, US: Hartley & Marks Publishers)
[29] Wisner B and Luce H R 1993 *GeoJ.* **30** (2) 127-140
[30] Ehrenfeld J R 2008 *Sustainability by design: A subversive strategy for transforming our consumer culture* (Connecticut, US: Yale University Press)
[31] Kemp R 1994 *Futures.* **26** (10) 1023-1046
[32] Robards M and Alessa L 2004 *Arctic.* **57** (4) 415-427
[33] Cova T J 1999 *Geogr. Inform. Syst.* **2** 845-858
[34] Tobin G A and Montz B E 2004 Natural hazards and technology: Vulnerability, risk, and community response in hazardous environments *Geography and Technology* eds S D Brunn, S L Cutter and J W Harrington JW (Netherlands: Springer) chapter 23 pp 547-570
[35] McEntire D A 2005 *Disaster. Prev. Manage.* **14** (2) 206-222
[36] Haque M S 1999 *Int. Polit. Sci. Rev.* **20** (2) 197-218
[37] Sianipar C P M, Dowaki K, Yudoko G and Adhiutama A 2013 *Eur. J. Sustain. Dev.* **2** (4) 1-18
[38] Gupta A K, Sinha R, Koradia D, Patel R, Parmar M, Rohit P, Patel H, Patel K, Chand V S, James T J, Chandan A, Patel M, Prakash T N and Vivekanandan P 2003 *Futures.* **35** 975-987
[39] Kaplinsky R 1990 *The economies of small: Appropriate technology in a changing world* (London, UK: Intermediate Technology Publications)
[40] Knack S and Keefer P 1997 *Q. J. Econ.* **112** (4) 1251-1288
[41] Kaplinsky R 2011 *Res. Policy.* **40** 193-203
[42] Clark N G 1985 *The political economy of science and technology* (Oxford, UK: Blackwell)
[43] Feng Y 2001 *Int. Stud. Quart.* **45** (2) 271-294
[44] Fields G S 2002 *Distribution and development: A new look at the developing world* (Boston,

IOP Conf. Series: Materials Science and Engineering **58** (2014) 012022 doi:10.1088/1757-899X/58/1/012022

US: MIT press)
[45] Ahmed S A, Diffenbaugh N S and Hertel T W 2009 *Environ. Res. Lett.* **4** (3) 034004
[46] Plenert G 1997 *Int. J. Technol. Manage.* **13** (4) 421-425
[47] Sianipar C P M, Yudoko G, Dowaki K and Adhiutama A 2013 *Sustainability.* **5** (8) 3382-3425
[48] Steenhuis H-J and de Bruijn E J 2001 *Technol. Soc.* **23** 551–562
[49] Dechezleprêtre A, Glachant M and Ménière Y 2009 *Energ. Policy.* **37** (2) 703-711
[50] Chaudhuri S, Jalan J and Suryahadi A 2002 *Department of Economics Discussion Papers* **0102-52** 1-36 (Columbia, US: Columbia University)
[51] Dasgupta A and Beard V A 2007 *Dev. Change.* **38** (2) 229-249
[52] Souza R, Bernatsky S, Reyes R and Jong K D 2007 *J. Trauma. Stress.* **20** (3) 263-269
[53] Garniati L, Owen A, Kruijsen J, Ishadamy Y and Wibisono I 2013 *Sustain. Cities. Soc.* doi: 10.1016/j.scs.2013.10.003
[54] Putranto K, Stewart D and Moore G 2003 *Technol. Soc.* **25** (1) 43-53
[55] Wie T K 2005 *J. Asia. Pac. Econ.* **10** (2) 214-236
[56] Sianipar C P M and Widaretna K 2012 *Procedia. Soc. Behav. Sci.* **52** 197-206
[57] Goodier R and Moseson A 2013 *Communities move to the center of the design process in a newly proposed methodology* Retrieved from http://goo.gl/Wm1SX2 on 2 October 2013
[58] Riley D 2008 *Engineering and social justice* (California, US: Morgan & Claypool)
[59] Yanu E P, Luthfiyanti R and Kumalasari R 2013 Implementasi Teknologi Tepat Guna untuk pemberdayaan masyarakat: Peluang, strategi dan tantangan *Seminar on Technology for the Poor (Jakarta, ID, 22 February 2013)*
[60] Juwaini A 2013 Peran riset dan teknologi dalam pemberdayaan masyarakat *Seminar on Technology for the Poor (Jakarta, ID, 22 February 2013)*

Synergization of silicone with developed crosslinking to soy-based polyurethane foam matrix

Flora Elvistia Firdaus

Jayabaya University, Jl. Pulomas Selatan Kav 23 Jakarta -13210 Indonesia

E-mail: flora_elvistia@yahoo.com

Abstract. Flexible polyurethane foam obtained from reaction of soybased polyol with TDI:MDI (80:20), and surfactant. The goal of this research is to determine the synergization effect of silicone with low molecular alcohols; methanol and ethylene glycol (EG) in soy-polyurethane formula on holding moisture of foams to density, foam solutions capacity, and cellular morphology. The optimized of polyol was achieved by ratio of epoxide/methanol 1:6 (mol/mol), and epoxide/EG 1:3 (mol/mol). It was found silicone surfactant can minimize solution absorbency in polyurethane foam matrix.

Keywords— soy polyurethane, surfactant, moisture, crosslinking

1. Introduction

Some research has explored and used various technological innovations as an effort to save environment as to respond the fears of supply of non-renewable natural resources to using renewables material [1]-[2]. Agricultural resources as one of many renewables has gaining attention of researchers for the environmental awareness and their potential to replace petroleum derivatives [3].

Polyurethanes (PUR) are the result of the exothermal reaction between a polyisocyanate and molecule containing two or more alcohols groups (-OH). Flexible properties owned by soft segmented with the percentages is higher than hard segment. It occured when ratio of polyol to isocyanate is greater than 1. Polyols will perform as soft segment while isocyanates forming hard segment. PUR products has range of wide spectrum from straight-chain polymers to thermosetting plastics. The structure and properties of polyurethane depends on density, hard and soft segments, and chemical compositions [4]-[11].

The flexible foam can be made as necessary by adding a chain extender, polyol derived from a short chain polyhydroxy defined as low molecular hydrocarbons, if not using chain extenders it can be added polyhydroxy initiator [12]. Chain extender are generally low molecular weight of reactant which produces hard segment in polyurethane, this believed as the result from an increased intermolecular association or bonding induced. The addition of small amount of water leads to the formation of urea linkages, which result in crosslinking. A component can be added to the prepolymer reaction to develop crosslinking, which can effect to the increasing the number of OH sites with which isocyanate can react and have reactive group [13].

The use of natural oils as raw materials for PUR production, multiple hydroxyl functionalities are required. By reacting epoxidised oils with low molecular weight or polyfunctional alcohols or acids. Unsaturated fatty acids in vegetable oil are plays an important role from the intermediates to final product. Fatty acid in soybean oil are constitute of palmitic acid 11%, stearic acid 4%, oleic acid 23%, linoleic acid 54%, and linolenic 8%. The combination of polyols, di isocyanate, and low molecular chain extender gives a multitude forms which suitable for extremely different practical applications [14]-[15].

Some advantages of using vegetable oils are having low toxicity, soluble, and high purity. Advantages possessed of using vegetable oils, because are easily to be reshaped and tailormade, which can be made according to market needs. Based on the previous research the optimized oxirane number was 6.7% [16]-[17]. The optimized formulation were then implemented to the proceeded steps.

This research was conducted an assessment of different polyurethane foam products synthesized from combinations of two low molecular weight alcohols; methanol represented as monol, and ethylene glycol represented as diol with a silicone surfactant inclusion. Furthermore, studied its role in holding the amount of solutions of an undisturbed foam which then verified to petrochemical-based polyurethane.

2. Materials and Methods

Soy polyol synthesis. Fifteen ml of soy-epoxide (oxirane number 6.7 mgr KOH/gr sample which processed from soybean oil with specification of kinematic viscosity 443.007 cps ; Iodine value 53.89 gr Iod/100 gram sample) was mixed to methanol with ratio (1:1; 1:3; 1:5 and 1:6) (mol/mol), and fifteen ml of soy-epoxide was also mixed to EG with ratio (1:1; 1:3; 1:5;1:6 and 1:10) (mol/mol) in two designated time; 1 hour and 2 hours for methanol and 1 hour, 2 hours, and 3 hours for EG with the concentration of acid catalyst 1% (v/v) and temperature 117^{0}C. The products obtained was neutralized, decantated, and filtered. The average number of –OH groups can lead to a certain controlled amount sites to develop crosslinking, which has the effect of increasing the number OH which then can react to isocyanate.

In making polyols from two route reactions; for methanol-based are namely as (M1;M2;M3 and M4) and EG based (EG1; EG2; EG3; EG4 and EG5). The products from each route are optimized before manufactured to polyurethanes namely as (PUR1) and (PUR2).

Overall, this determination is to identify which of these route of reaction will result the best property to PUR products. The expected oxirane number in polyol synthesis is the lowest among other compositions, which were conducted in triplo in a stable average number.

PUR synthesis. PUR were made by mixing soy-polyol, TDI: MDI (80:20), surfactants, and blowing agent, distilled water. The mixture is then poured into the glass mold, after a couple of minutes being completely dried and safe to be appointed from the molding.

2.1.Method of Analysis

- Oxirane number: specify a group of oxirane oxygen obtained from the titration using HBr in glacial acetic acid.

- Foam Density test: the sample that had been prepared in a particular dimension was weighed using an analytical balance, and then calculate the volume of dimension.

$$\text{Density} = M/W \qquad (1)$$

where

M: mass specimen (gram)

V: volume of specimen (cubic cm)

- Foam water capacity: measures the amount of solution held in foam matrix to specimen not less than 10 cc.

$$\text{Water Capacity (\%)} = [W_w/w_d]*100 \qquad (2)$$

While, oxirane number of polyol EG-based for 1 hour reaction was in the range 0.09-0.2 % with oxirane reduction 90-98%; for 2 hour reaction was 0.08-0.3% with oxirane reduction 96-99%, and for 3 hour reaction was 0.1-0.7% with oxirane reduction 90-99% as can be seen in figure 2.

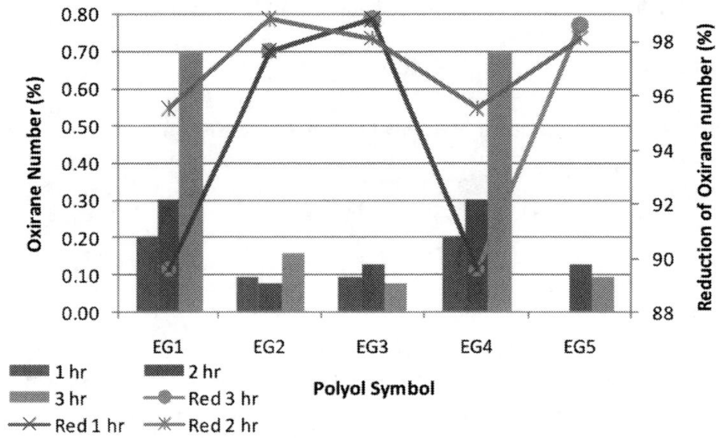

Figure 2. Soy-polyol synthesized from soy-epoxide hydroxylation with EG as a source for develop cross linkers. It was made in five formula. The graph illustrates relationship between epoxide oxirane number referred to optimized epoxide oxirane number 6.7% to the percentages reduction of oxirane number after soy-polyol formation

The high percentage reduction of oxirane number of EG compared to methanol, does not concerted to the high hydroxyl number. The hydroxyl number of EG was 309 mgr KOH/ gr sample as can be seen in figure 3. This indicates the developed cross link of methanol-based is higher than EG-based, and caused to the high –OH sites to methanol-based.

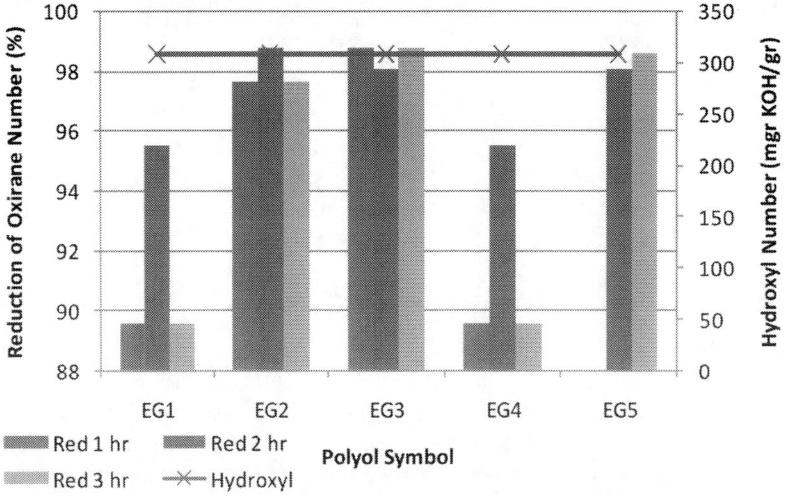

Figure 3. Soy-polyol synthesized from soy-epoxide hydroxylation with EG as a source for develop cross linkers. It was made in five formula. The graph illustrates relationship between the (%) reduction of soy-polyol oxirane to the formation of -OH sites in soy-polyol determined by hydroxyl number

Using EG in the second stage of the method it is possible to obtain soy-based polyols with primary and secondary hydroxyl groups. The aliphatic β-OH in EG has effected to hydroxyl number of polyol, referred to steric hindrance during hydroxylation reaction which in methanol only have aliphatic α-OH.

B. PUR Characterization
Density

The 1 ml amount of silicone in PUR formulation is seems not to be significantly effected to apparent density of the foam. The volume molding relates to the space free rising of the foam which caused to lowering density. Overall, foams molding with volume 500 mLgives density and density distribution are lower than volume 250 mL.

Table 1 The Apparent Density of dry molded soy-PUR

Type of PUR	Density (gr/cm3)	Density Distribution (%)
Without Silicone		
Methanol (250)	0.12	74
EG (250)	0.10	15
Synthetic (250)	0.13	33
With Silicone		
Methanol (250)	0.07	16
Methanol (500)	0.06	11
EG (250)	0.07	11
EG (500)	0.06	17

*Amount of silicone is 1 ml
The molding were designed in 250 mL and 500 mL

Solution Capacity to PUR matrix

The absorbency of foam matrix its seems does relate to density distributions. The high density distribution in the foam matrix more distance between the open cellular. Water capacity for 250 mL molded PUR is higher than 500 mL molded PUR. The addition of silicone in PUR formula has made to the low foam matrix water capacity.

The immersion of etanol (Et-OH) 96% into foam matrix has made a significant effect to using silicone and without silicone in PUR formulation; with silicone the foam matrix has the capacity to hold the Et-OH solutions over 100%. It was found in the general pore diameter for methanol-based PUR was 7.8 mm and EG-based PUR was 5 mm.

Table 2. PUR foam matrix to the capacity of holding solutions

Type of PUR	Density (gr/cm3)	Water in Foam (%)	Foam Water Capacity (%)	Et-OH 96% in Foam (%)	Foam Et-OH 96% Capacity (%)
Without Silicone					
Methanol (250)	0.12	87	231	0.20	0.09
EG (250)	0.10	89	189	0.19	0.09
Synthetic (250)	0.13	(-)	(-)	0.06	0.06
with Silicone					
Methanol (250)	0.07	7	107	0.07	100.07
Methanol (500)	0.06	14	114	0.11	100.11
EG (250)	0.07	17	117	0.07	100.07
EG (500)	0.06	18	118	0.15	100.15

*The conditions was made for undisturbed foam matrix
Solutions used; 4 drops of water and 4 drops of Et-OH 96% in the foam dimension (1cmx1cmx1cm)

The justification of the result above confirmed by P foam cell morphology using SEM (figure 5). Visible cavities (void) generated by the methanol-based PUR2 is larger and comparable to EG-based PUR. It was found in the general pore diameter for methanol-based PUR was 7.8 mm and EG-based PUR was 5 mm.

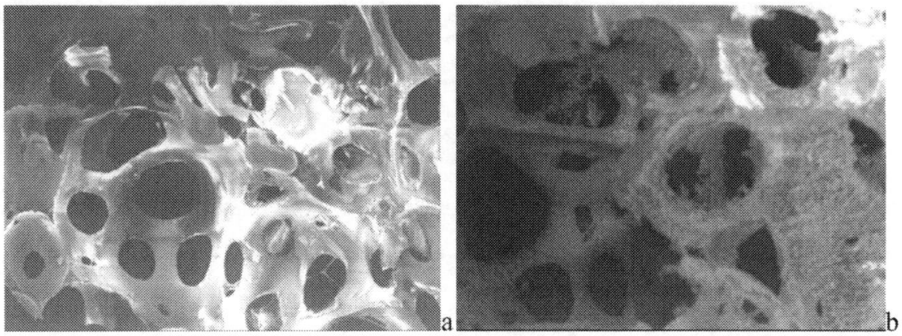

Figure 4. Cell Morphology of a. Methanol-based PUR 40xμm and
b. EG-based PUR 45x 500 μm

Conclusion

The present results confirmed the high percentages reduction of epoxide oxirane number occurred during polyol synthesis does not brought to the conclusion to the high hydroxyl number. The developed cross linking existence from low molecular alcohol are depends on the –OH group location during the hydroxylation reactions.

According to the research founding, silicone surfactant are effective in lowering foam imbibe to water to either methanol-based PUR or EG-based PUR, but this does not valid to Et-OH 96%.

Acknowledgement

The authors gratefully acknowledge to the support of the Indonesia Department of Higher Education (Dikti-Kementerian Pendidikan dan Kebudayaan Republik Indonesia) as funding the research.

References

[1] Narine, Kong, Bouzidi dan Sporns, Physical properties of polyurethane produsced from polyol seed oils; elastomers. J Amer Oil Chem Soc 84: 2007, pp. 55-63
[2] Guidoti *et.al*, Heterogenous catalytic epoxidation of fatty acid nd methl esters on titanium grafted silicas, Green Chem, 5, 2003, pp. 421-424
[3] Chuajulit S, Sangpakdee T, Saravari O (2007), J Met Mater Miner 17(1), pp.17-23.
[4] Kong and Narine 2007. Physical properties of polyurethane plastic sheets produced from polyols from canola oil biomacromolecules 8; 2007, pp. 2203-2209
[5] Hanson and Larock, Soybean oil divinilbenzene thermosetting polymers: synthesis, structure, properties and their realtionships. Pollymer 42(4):2001, pp. 1567-1579
[6] Petrovic, Cevalos, Javni, Schaefer, and Justice 2005. Soy oil based segmented polyurethanes. J Polym Phys 43(22):2005, pp. 3178-3190
[7] Petrovic, Zhang, and Javni 2005. Structure and properties of polyurethane prepared from triglycerides polyols by ozonolysis. Biomacromolecules 6(2):2005, pp. 713-719
[8] Zlatanic, Petrovic, and Dusek, Structure and properties of triolein based polyurethane networks, Biomacromolecules, 3: 2002, pp. 1048-1056
[9] Lligadas, Ronda, Galia and Cadiz, Polyurethane networks from fatty acid based aromatic triols: synthesis and characterization. Biomacromolecules 8:2007, 1858-1864
[10] Andjelkovic and Larock, Novel rubbers from cationic copolymerization of soybean oils and dicyclopentadiene 1: synthesis and characterization. Biomacromolecules 7:2006, pp. 927-936
[11] Guner, Yagci, and Erciyes 2006. Polymers from triglyceride oils. Prog Polym Sci 31: 2006, pp. 633-670
[12] Aneja,Structure Property Relationships of flexible Polyurethane Foams. Dissertation. 2000, Virginia Polytechnic Institute.
[13] Thomson T., Design and Applications of Hydrophilic Polyurethanes Technomics Publishing Company, Inc., USA, 2000, p.7-8
[14] Wang, T.L, Hsiech, Polym J, 29 1996, p.839
[15] Bauman, H., Bühler,M., Fochen, H., Hirisnger, F. Zoebelien Angew, H., Chem. Int. Ed Engl., 27, 1998, pp.41-62
[16] Flora, F.E, Optimization of Soy Epoxide Hydroxylation to Properties of Prepolymer polyurethane, Proceeding International Conference on Applied Science Engineering and Information Technology, ISSN 2010-3778, Waset v81 2011, Singapore, p. 189
[17] Flora, F.E, Silicone on Blending Vegetal Petrochemical Based Polyurethane, Proceeding International Conference on Applied Chemistry, ISSN 2010-3778, Waset 65, 2010, Tokyo, p.350
[18] Patent US 2008293913 A1

MOIME 2014 IOP Publishing
IOP Conf. Series: Materials Science and Engineering **58** (2014) 012024 doi:10.1088/1757-899X/58/1/012024

A two-level model of rise time in quantum cascade laser materials applied to 5 micron, 9 micron and terahertz-range wavelengths

J F Webb[1], K S C Yong[2] and M K Haldar[2]

[1] Department of Mechanical, Materials and Manufacturing Engineering, The University of Nottingham Malaysia Campus, Jalan Broga, 43500 Semenyih, Selangor Darul Ehasn, Malaysia
[2] Faculty of Engineering, Computing and Science, Swinburne University of Technology, Sarawak Campus, Jalan Simpang Tiga, 93350 Kuching, Sarawak, Malaysia

E-mail: `jeffwebb@physics.org`

Abstract. An equivalent circuit simulation of a two-level rate equation model for quantum cascade laser (QCL) materials is used to study the turn on delay and rise time for three QCLs with 5 micron, 9 micron and terahertz-range wavelengths. In order to do this it is necessary that the model can deal with large signal responses and not be restricted to small signal responses; the model used here is capable of this. The effect of varying some of the characteristic times in the model is also investigated. The comparison of the terahertz wave QCL with the others is particularly important given the increased interest in terahertz sources which have a large range of important applications, such as in medical imaging.

1. Introduction

Quantum cascade lasers (QCLs) [1] have many applications. Of particular relevance to the work here are applications in communication systems. For free space communications[2] measurements have shown that there are two windows, 3 to 5 μm and 8 to 13 μm, in which the atmosphere is relatively transparent compared to near infrared wavelengths. At high altitudes terahertz-range waves can be used for communications. This range has many other applications, such as in medical imaging and spectroscopy. Therefore in this paper, we consider three QCLs with the emission wavelengths of 5 μm [3], 9 μm [4] and, for the terahertz-range, 103 μm [5].

A lot of numerical[6–9] and analytical[3, 4, 10] work has been published on the modulation behaviour of QCLs. But this work has been concerned with steady-state and small-signal responses. Alone this is not enough to adequately describe the range of the modulation behaviour; the large signal response for pulse code modulation is also important. In order to deal with this we employ an equivalent circuit model[6] of a QCL based on improved two level rate equations[3]. The equivalent circuit (Fig. 1) is relatively straightforward can easily deal with both small and large signal responses which is convenient for studying the turn-on delay and rise time of the QCLs considered here.

Under the following assumptions [3]:

(i) all quantum wells and injectors are identical and hence the gains of the wells are the same,
(ii) confinement factors are identical,

- Hydroxyl number (OH number): a number arising from a wet analytical method; milligrams of KOH equivalent to the hydroxyl content in one gram of polyol hydroxyl compound.

  OH number = 56.1 x 1000/equivalent weight (3)

- Density Distribution (%) = [max. density- min. density/average density]x 100 (4)

- Solution Percentages (%) = $[W_w\text{-}W_d/W_d]$ *100 (5)

- % oxirane reduction = 6.7% (optimized oxirane)-Ox_c / 6.7% x 100 (6)

where:

Ox_c ; the current oxirane number

3.Result and Discussion

A. Synthesis of Polyol.

Hydroxylation groups have been introduced from two-step synthesis which involving epoxidation of unsaturated sites with a mixed of peroxide with acetic acid; followed by epoxy ring opening with mono functional and difunctional alcohol.

The oxirane number occured from polyol methanol-based for 1 hour reaction was in the range of 0.11-2.47% with the range percentages of oxirane number reduction was 63-95% ; the oxirane number for 2 hour reaction was 0.14-1.95 % with range percentages of oxirane reduction was 71-90% but it has reached hydroxyl number to 579 mgr KOH/gr sample as compared to palm oil-based polyol with same method is 70-130 mgr KOH/gr [18]. The graph can be seen in figure 1.

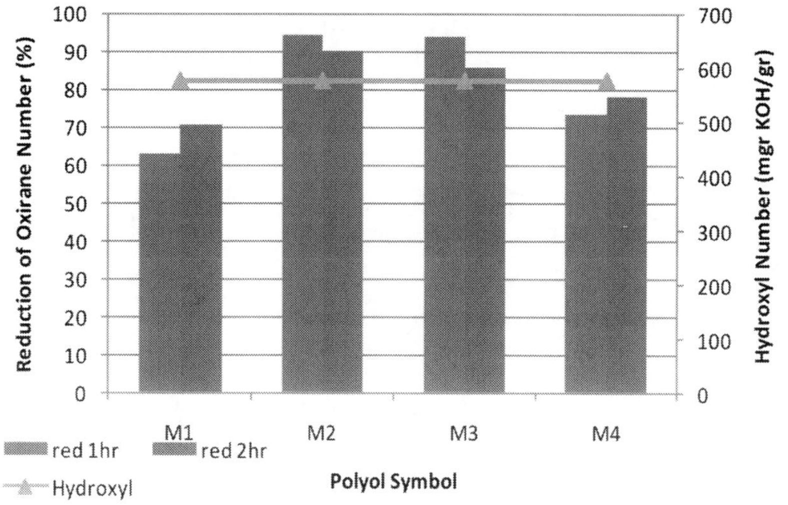

Figure1. Soy-polyol synthesized from soy-epoxide hydroxylation with methanol as a source for develop cross linkers. It was made in four formula. The graph illustrates relationship between the percentages reduction of soy-polyol oxirane number referred to optimized epoxide oxirane number 6.7% to the formation of -OH sites in soy-polyol determined by hydroxyl number

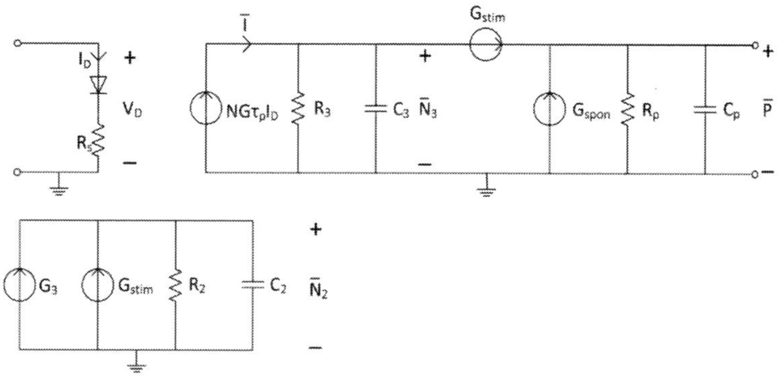

Figure 1. Equivalent circuit for a QCL modelled from normalized two-level rate equations. The normalized quantities are $\bar{I} = NG\tau_P I$, $\bar{N_3} = NG\tau_P N_3$, $\bar{N_2} = NG\tau_P N_2$ and $\bar{P} = P/K$, with $K = 1/[G\tau_3(1+\tau_{21}/\tau_{31})]$. Also, $V_D = nV_T \ln(I/I_S + 1) + I_D R_S$, n is a diode ideality factor, $V_T = kT/e$, k is Boltzmann's constant, e is the electronic charge magnitude, $G_{\text{stim}} = eKG(\bar{N_3} - \bar{N_2})\bar{P}$, $G_{\text{spon}} = e\beta\bar{N_3}/\tau_{sp}$, $G_3 = e(1/\tau_{32} + 1/\tau_{sp})\bar{N_3}$, $R_3 = \tau_3/e$, $R_2 = \tau_{21}/e$, $C_3 = C_2 = e$, and $C_P = eKG\tau_P$, $R_P = 1/(eKG)$.

(iii) removal rates for electrons are fast,

(iv) the coupling of the spontaneous emission to the lasing mode can be ignored,

the two level rate equations of a QCL including the coupling of the spontaneous emission to the lasing mode can be written as [3]:

$$\frac{dN_3}{dt} = \frac{I}{e} - \frac{N_3}{\tau_3} - G(N_3 - N_2)P \tag{1}$$

$$\frac{dN_2}{dt} = \left(\frac{1}{\tau_{32}} + \frac{1}{\tau_{sp}}\right)N_3 - \frac{N_2}{\tau_{21}} + G(N_3 - N_2)P \tag{2}$$

$$\frac{dP}{dt} = NG(N_3 - N_2)P + N\beta\frac{N_3}{\tau_{sp}} - \frac{P}{\tau_p} \tag{3}$$

where N_2 and N_3 are the electron numbers in levels 2 and 3 respectively; I is the injected current; G is the gain coefficient per stage, which is the product of the confinement factor and the gain due to stimulated emission; τ_{31}, τ_{32} and τ_{21} are the lifetimes representing the transitions from levels 3 to 1, 3 to 2 and 2 to 1 respectively; τ_{sp} is the spontaneous lifetime between level 3 and 2; τ_P is the photon lifetime; P is the photon number in the cavity; N is the number of stages; β is the fraction of spontaneous emission entering the lasing mode; and t is time. τ_3 is given by $1/\tau_3 = 1/\tau_{31} + 1/\tau_{32} + 1/\tau_{sp}$. As discussed in Ref. [6], if Eqs. (1) to (3) are used to derive the circuit simulator it will fail to converge. This problem is solved by employing a systematic normalization procedure given in Ref.[6].

2. Results and Discussion

Turn-on delay and rise time of the three QCLs at the wavelengths 5 μm, 9 μm, and 103 μm are obtained through numerical simulation using the equivalent circuit model outlined above[6]. The parameter value for each QCL is given in Table 1.

Turn-on delay is defined as the time taken for the photon number to reach 10% of its steady value upon switching on. Rise time is defined as the time taken for the photon number to rise

Parameters			
Wavelength (μm)	5	9	103
Number of gain stages, N	25	48	30
Total confinement factor, Γ_N	0.5	0.32	0.27
Cavity width, W (μm)	11.7	34	80
Cavity length, L (mm)	3	1	3
Facet reflectivity, R	0.27	0.29	0.29
Mode-group index, n_g	3.4	3.27	3.3
Cavity internal loss, α_i (cm^{-1})	11	20	24
Differential gain, a (cm)	4×10^{-9}	4×10^{-9}	5×10^{-8}
Gain coefficient per stage, $G = a\frac{\Gamma_N}{N}\frac{c}{n_g LW}$ (s^{-1})	2×10^3	0.7×10^3	1.7×10^3
Spontaneous emission factor, β	2×10^{-3}	2×10^{-3}	2×10^{-3}
τ_3 (ps)	1.3	1.4	1.1
τ_{32} (ps)	2.1	2.1	2
τ_{31} (ps)	3.4	4.2	2.4
τ_{21} (ps)	0.5	0.3	0.3
τ_p (ps)	7.38	3.36	3.7
τ_{sp} (ns)	6.06	35.5	7
Threshold current, I_{th} (A)	0.44	1.11	0.91

Table 1. Parameter values for the 5 μm, 9 μm and 103 μm QCLs

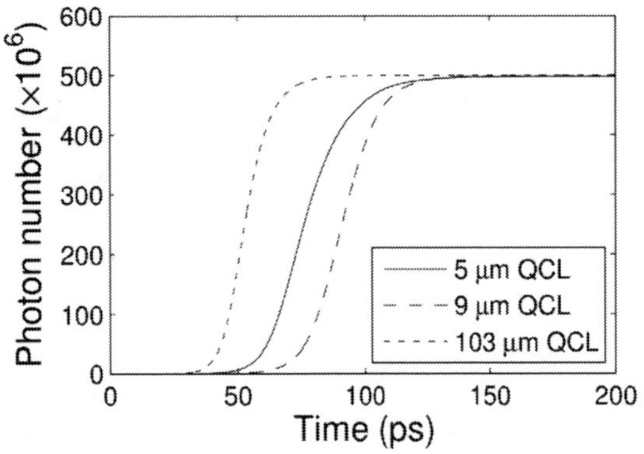

Figure 2. Turn-on response for 5 μm, 9 μm and 103 μm QCLs. Value of injected current has been carefully selected to give the same photon number output. Current is applied at $t = 0$.

from 10% to 90% of its steady value. Here, we have investigated the transient response for the three QCLs for a fixed photon number. This photon number is set at 500×10^6.

Figure 2 shows 103 μm QCL has the shortest turn-on delay of 43 ps and followed by 5 μm and 9 μm QCLs with turn-on delay of 61 ps and 76 ps respectively. As for rise time, 103 μm QCL has the smallest value of 22 ps and followed by 9 μm and 5 μm QCLs with rise time of 32 ps and 37 ps respectively. To fully understand why the response for the three QCLs differ as such, we observe how a single parameter affect the response. Two parameters chosen in this paper are τ_p and τ_{sp} as these parameters have significant difference between the three QCLs. Only one parameter will have its value changed each time while the rest of the parameters value will remains the same. The results are plotted in Figure 3.

Figure 3 shows that τ_p affects both the turn-on delay and rise time of QCLs. Shorter photon

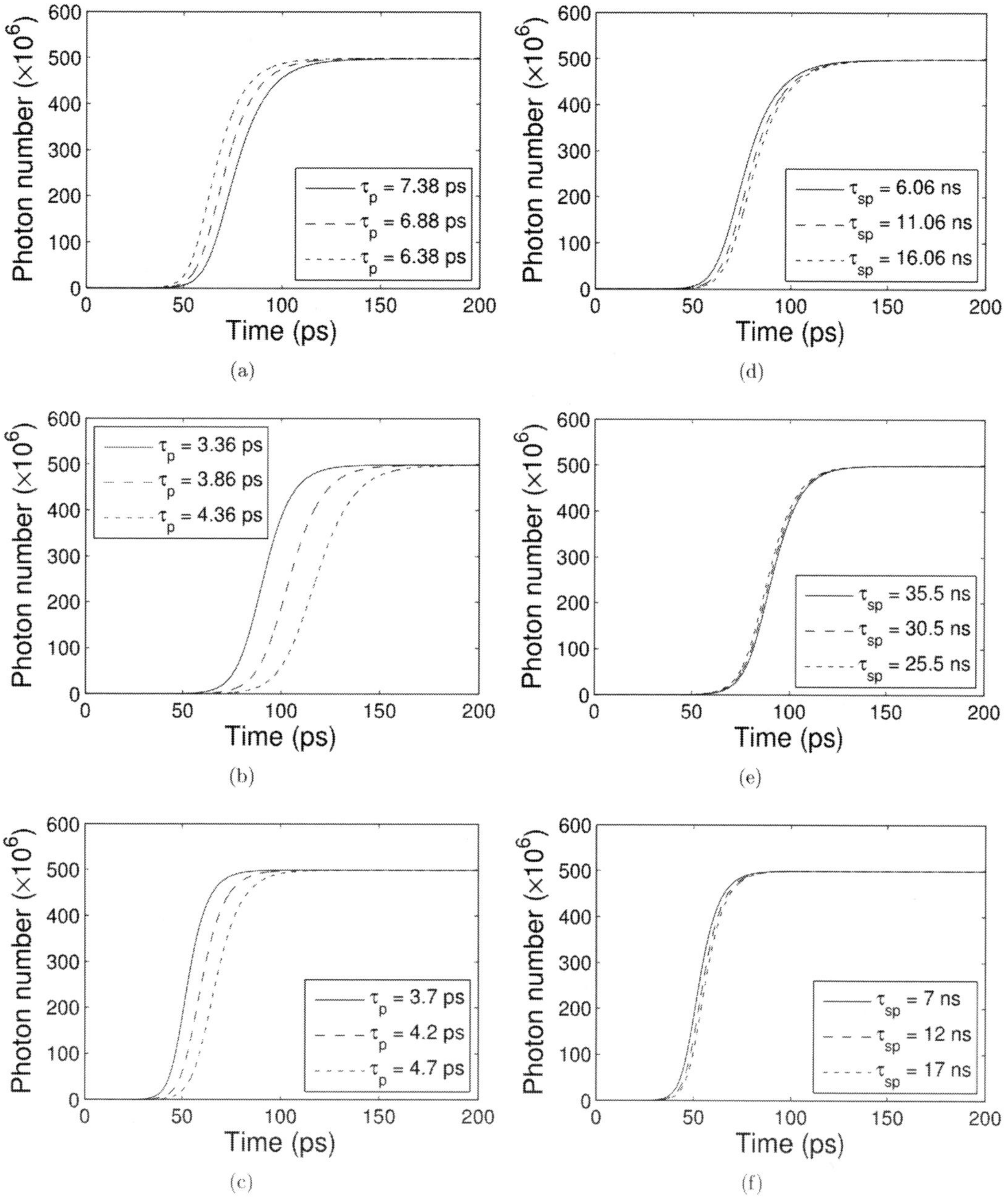

Figure 3. (First column)Turn-on response for (a)5 μm (b) 9 μm and (c) 103 μm QCL with different value of τ_p. (Second column) Turn-on response for (d)5 μm (e) 9 μm and (f) 103 μm QCL with different value of τ_{sp}. Value of injected current has been carefully selected to give the same photon number output. Current is applied at $t = 0$.

lifetime contributes to lower turn-on delay and rise time. As for τ_{sp}, this parameter affects the turn-on delay of QCL only. Shorter radiative spontaneous lifetime results in lower turn-on delay. Hence, we can deduce that the 103 μm QCL has the least turn-on delay and rise time

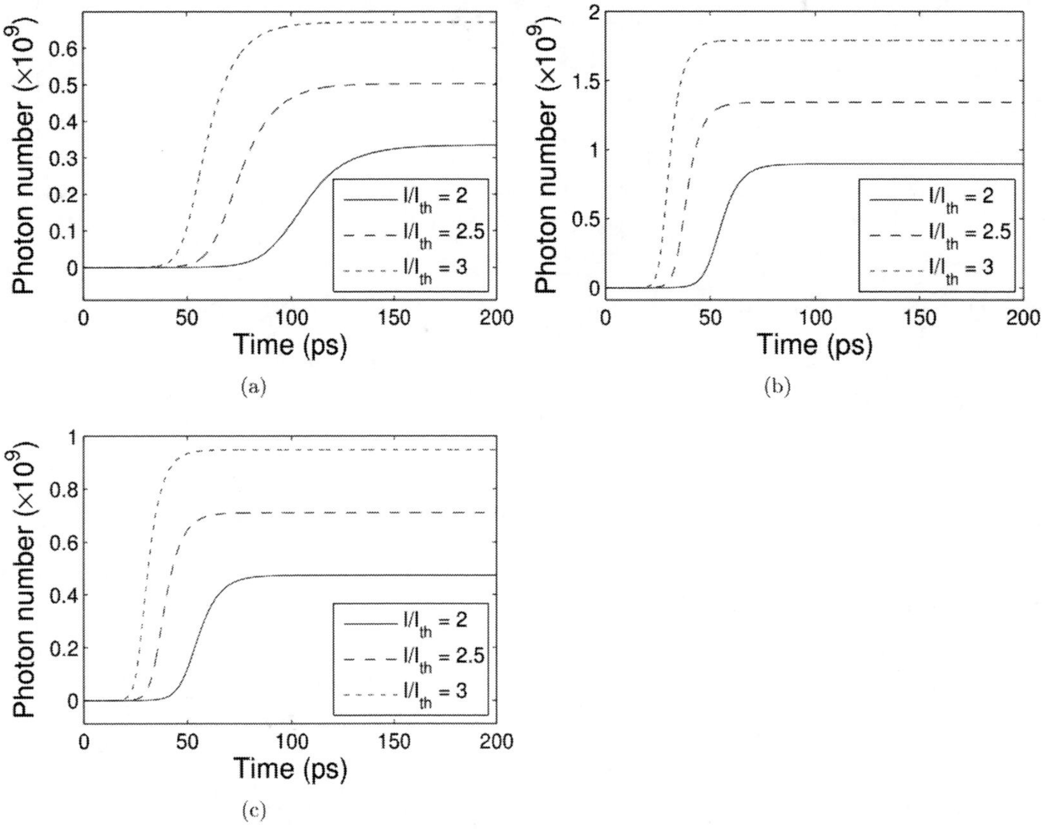

Figure 4. Turn-on response for (a)5 μm (b) 9 μm and (c) 103 μm QCL with different values of injected current.

partly because it has relatively low values of τ_p and τ_{sp} compared to the 5 μm and 9 μm QCLs. Parameters that affect turn-on delay and rise time are not limited to τ_p and τ_{sp}; other parameters will also affect the transient response but will not be discussed here.

Now, we investigate the effect of injected current on turn-on delay and rise time. The transient response of different level of injected current for the three QCLs is shown in Figure 4. Turn-on delay decreases with increasing injected current for all three QCLs. This is consistent with the finding of Hamadou et al.[4]. We have also observed that a higher level of injected current results in a lower rise time.

3. Conclusion

In conclusion, we have shown in this paper that the turn-on delay and rise time of quantum cascade lasers at three different wavelengths depends on photon lifetime, radiative spontaneous lifetime and injected current value. Currently, we are working on analytical expression for turn-on delay, rise time and fall time of pulse-modulated QCLs to give a better understanding of how various parameters affect the QCL transient response.

[1] Capasso F, Paiella R, Martini R, Colombelli R, Gmachl C, Myers T, Taubman M, Williams R, Bethea C, Unterrainer K, Hwang H, Sivco D, Cho A, Sergent A, Liu H and Whittaker E 2002 *J. Quant. Eletron.* **38** 511–532

[2] Taslakov M, Simeonov V and van den Bergh H 2008 *Free-space Laser Communication Technologies XX* (*Proc. SPIE* vol 6877) ed Mecherle S (Bellingham: SPIE) p 68770F
[3] Webb J F and Haldar M K 2012 *J. App. Phys.* **111** 093116–1–093116–6
[4] Hamadou A, Lamari S and Thobel J L 2009 *J. App. Phys* **105** 0139116–1 to 0139116–6
[5] Petitjean Y, Destic F, Mollier J C and Sirtori C 2011 *IEEE J. Sel. Topics Quantum Electron.* **17** 22–29
[6] Yong K S C, Haldar M K and Webb J F 2013 *J. Infrared Milli. Terahz. Waves* **34** 586–597
[7] Chen G, Fan G and Li S 2008 *Opt. Quantum. Electron.* **40** 645–653
[8] Petitjean Y, Destic F, Mollier J C, Barbieri S and Sirtori C 2009 *34th IRMMW-THz* vol 1 & 2 (New York: IEEE) pp 270–271
[9] Yang J, Zhou W, Li X, Wu J and Xia S J 2012 *Adv. Mat. Res.* **490** 61–65
[10] Haldar M K A simplified analysis of direct intensity modulation of quantum cascade lasers *J. Quant. Eletron.* **41** 1349–1355

AUTHOR INDEX

Institute of Physics
Dirac House, Temple Back
Bristol BS1 6BE UK

CURRAN ASSOCIATES INC
proceedings
.com